Austro-Hungarian Fighter Units of WWI

Volume 1 | Fliks 68/J, 72/J, & 74/J

Zoltán Czirók

Acknowledgements
Color aircraft profiles © Alexandr Kazakov.

For our aviation books, please see our website at: www.aeronautbooks.com.

I am looking for photographs of the less well-known German aircraft of WWI to complete this series. For questions or to help with photographs you may contact me at jherris@me.com.

Interested in WWI aviation? Join The League of WWI Aviation Historians (**www.overthefront.com**), Cross & Cockade International (**www.crossandcockade.com**), and Das Propellerblatt (**www.propellerblatt.de**).

ISBN: 978-1-953201-39-3

Design and layout: Jack Herris
Cover design: Aaron Weaver
Digital photo editing: Jack Herris

Aeronaut Books
Books for Enthusiasts by Enthusiasts
www.aeronautbooks.com

Table of Contents

Preface and Acknowledgements

It was more than two decades ago when I became fascinated with the stories of the Austro-Hungarian Monarchy's aviation troops and, as I became more and more immersed in the subject, this fascination only grew as the years went on. I can consider myself lucky, as this passion took shape at a time when several people turned to World War I aviation and, to the delight of many of us, they developed an interest. It was a particular pleasure for me to see the attention given to subjects such as the air wars on the southwest and eastern fronts, as well.

During my own research, I also had the opportunity to follow the path paved by many great aviation historians and their work and experience made it much easier to follow the maze of contemporary events. Some of them have helped me indirectly, "just" through their writings, but I am now in direct contact with many of the great experts and we try to support each other in the work and fun that this subject entails.

Early on in the Austrian War Archives (Österreichisches Kriegsarchiv), I had to face the fact that the *Luftfahrtarchiv* is not a separate collection of documents from the aviation troops (*k.u.k. Luftfahrtruppen*) in the traditional way, but principally from other troops and operational commands or departments, and these do not necessarily form a coherent whole. For example, there are no record books or war diaries of the aviation units, and there are serious gaps between the available documents, which in many cases are insurmountable. This is probably the reason why the books and studies of unit histories are almost completely lacking and why so few works have attempted to present the complete history of a squadron and its actions. Fortunately, however, certain aspects of the air war have been dealt with, such as air victories and losses, and we cannot pass the subject by without mentioning the pioneering work of Karl Meindl, who essentially provided an excellent basis for all subsequent researchers with his life's work.

These experiences and motivations led me to start working on and presenting the history of the various Austro-Hungarian flying units, or at least to attempt to do so. I believe that after more than a century, the time has come for this project. This planned series examines the history of Austro-Hungarian fighter units, not only the fighter squadrons but also the provisional fighter units, from the famous "ace" squadrons to the little or not at all known ones. Actually, the reader will read about the lesser-known units before the famous ones, and hopefully also the naval fighter units. I think it is important that the spotlight should not only be on the victorious air battles, but any NCO who flew even a single mission should also have a place in the story, even if he did not make a name for himself during the war.

It may seem strange and unusual in the way I have approached the history of a particular squadron. The reader will not find any in-depth analysis of the field operations, the background processes, the difficulties faced by the Austro-Hungarian army and the aviation troops, etc. The focus is on the operational deployment of the units. It is a kind of extended squadron record book, listing the missions in as much detail as possible, quantifying the performance of the pilots and, where possible, identifying the participants in combats. The text provides a highly detailed description of squadron missions using combat reports, first-hand accounts, and memoirs.

I hope this style and approach will appeal to the enthusiasts who are interested in the history of Austro-Hungarian aviation, and they will find it interesting and useful and at least partly gap-filling.

This volume covers the fate of three squadrons, the most recently formed fighter squadrons. The careers of two of them were very short, but this is not surprising, since in the spring and summer of 1918 the Austro-Hungarian army was already at a disadvantage in every aspect against the enemy, and the same was true of the air troops. Still, the units, bleeding from many wounds and suffering heavy losses, did their best and took up the fight against their opponents. They were mostly outnumbered and outclassed. This series and the first volume of this series, which the reader holds in his hands, are intended to commemorate Austro-Hungarian aviators in their own way.

Certainly, no research can be considered complete. My work is based on the sources and literature available so far, and I hope that further data will emerge in the future that will expand the information we have on these squadrons.

There are many people who have made this volume and this series possible.

First of all I would like to thank my wife, Christina, who has always understood how much the research and publication of this subject meant to me and has faithfully supported me throughout my research.

I could never have carried out my research at a significant pace and so carefully without the dedicated and continuous help of the former and current member of the Hungarian Archive Delegation in the Austrian War Archives: György Domokos, Gábor Kiss, Balázs Lázár, Ferenc Lenkefi, József Solymosi, and Norbert Számvéber. I am especially indebted to all of them.

I am also very obliged for the kind staff of all the archives where I did any research.

I have been so blessed by the support of numerous aviation historians: Boris Ciglić, Roberto Gentilli, Antonio Iozzi, Árpád Kajon, Andy Kemp, Giorgio Marinello, Paolo Miana, Jiři Rajlich, Christian Reiterer, Petr Aharon Tesař, Vittotio Valentino, Paolo Varriale, Aaron Weaver, Marcus Zelezny. Very special thanks to all of them for their assistance and their generosity.

I am particularly grateful to Aleksandr Kazakov for his amazingly beautiful color profiles.

Finally, I would especially like to thank my publisher, Jack Herris, for having enough faith in me and my project to publish this series. It is a great honor that Aeronaut Books has published my work. I would like to express my gratitude to Eugene Ushakov whose idea was the common work with Jack, and I can always enjoy his encouraging faith in me.

Zoltán Czirók

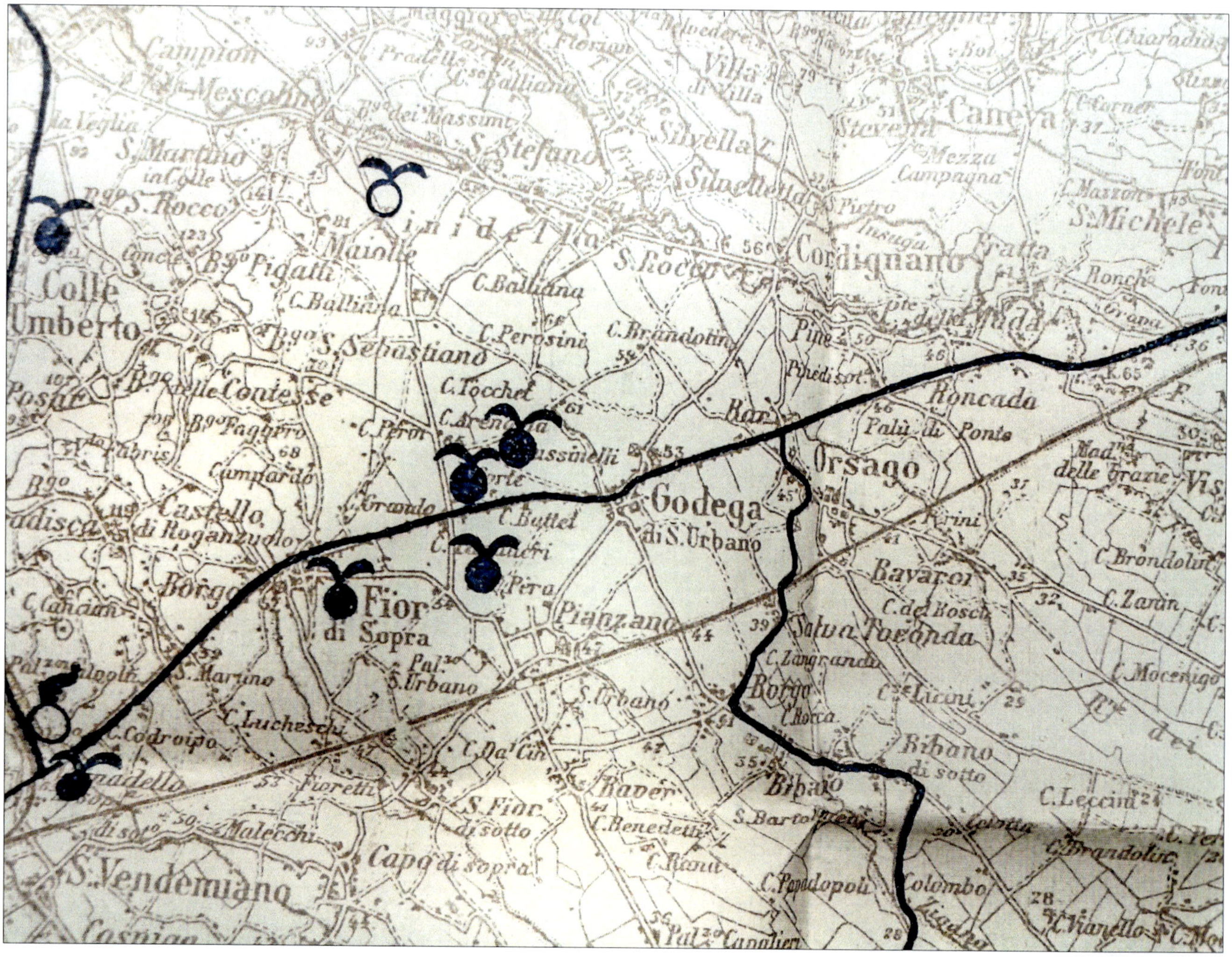

Above: Map of the area of San Fior – Pianzano – Godega (Paolo Miana)

AIRFIELDS OF THE SQUADRONS: GODEGA DI SANT'URBANO AND SAN FIOR DI SOPRA

The flat fields of Veneto, usually seeded with corn or wheat, were easily converted to airfields, and the presence of large brick farms or landowners mansions provided suitable accommodation for squadron personnel and officers. The area where the newest fighter squadrons arrived was an important and central part behind the frontline in the neighbourhood of Conegliano. There were three airfields close to each other: San Fior di Sopra, Pianzano, and Godega di Sant'Urbano.

The last mentioned airfield was built between 1916 and 1917 as an emergency airfield on the route of Turin-Pordenone and used for the delivery of aircraft to the front. After the breakthrough of Caporetto, Austro-Hungarian and German squadrons arrived at Godega: Flik 56/J on 24th November 1917 (it was stationed here until the end of December), FA 14 and FA(A) 232 in the first half of December 1917, and Flik 53/D which was ordered to move there on 4th January 1918.

The airfield was divided into two areas, actually, there were two airfields next to each other. One of them was west of Godega, north of the road San Fior di Sopra–Godega di Sant'Urbano (named Balliana in the Italian intelligence reports), and the other was northwest of the town and north-northeast of the road (Tocchet). According to an order dated 15th January 1918, the squadrons were stationed at Godega or were ordered to relocate there: Flik 4/D, 28/D, 57/F on the airfield west of Godega, and FA 14, FA(A) 219, 232 on the one

Above: Aerial photo of the airfield of Godega di Sant'Urbano on 5th December 1917. (Paolo Miana)

Above: Aerial Aerial view of the airfields of Pianzano and Godega on 18th March 1918. Note the black arrow showing the the eastern direction. (Paolo Miana)

northwest of the town. However, Flik 4 never arrived, Flik 70/D served instead.

In March 1918, ten canvas hangars were erected on the northern field and 11 on the southern one. Italian airships bombed the airfield(s) at least three times in February-March and it did not stop in the last half-year of the war either, suffering from several bombing attacks by Italian aircraft and airships.

The sketch of Flik 68/J shows that the Austro-Hungarian fighter squadron used the southern airfield (while the official station of the unit was in Colle Umberto). Duringthe period of May-July 1918, 13-15, hangar tents were reported by the Italians. In the monthly reports of the squadron, however, the building of five wooden hangars and an officer barracks were mentioned, which were completed by July.

The base of Fliks 72/J and 74/J was at San Fior di Sopra, located on the eastern-southeastern edge of the town, along the south of the main road to Godega di Sant'Urbano. It was used first by German fighter units from December 1917, which was after German and Austro-Hungarian troops moved forward to the Piave. Between the middle of December 1917 and the first days of January 1918, all three German Jastas (Jasta 1,

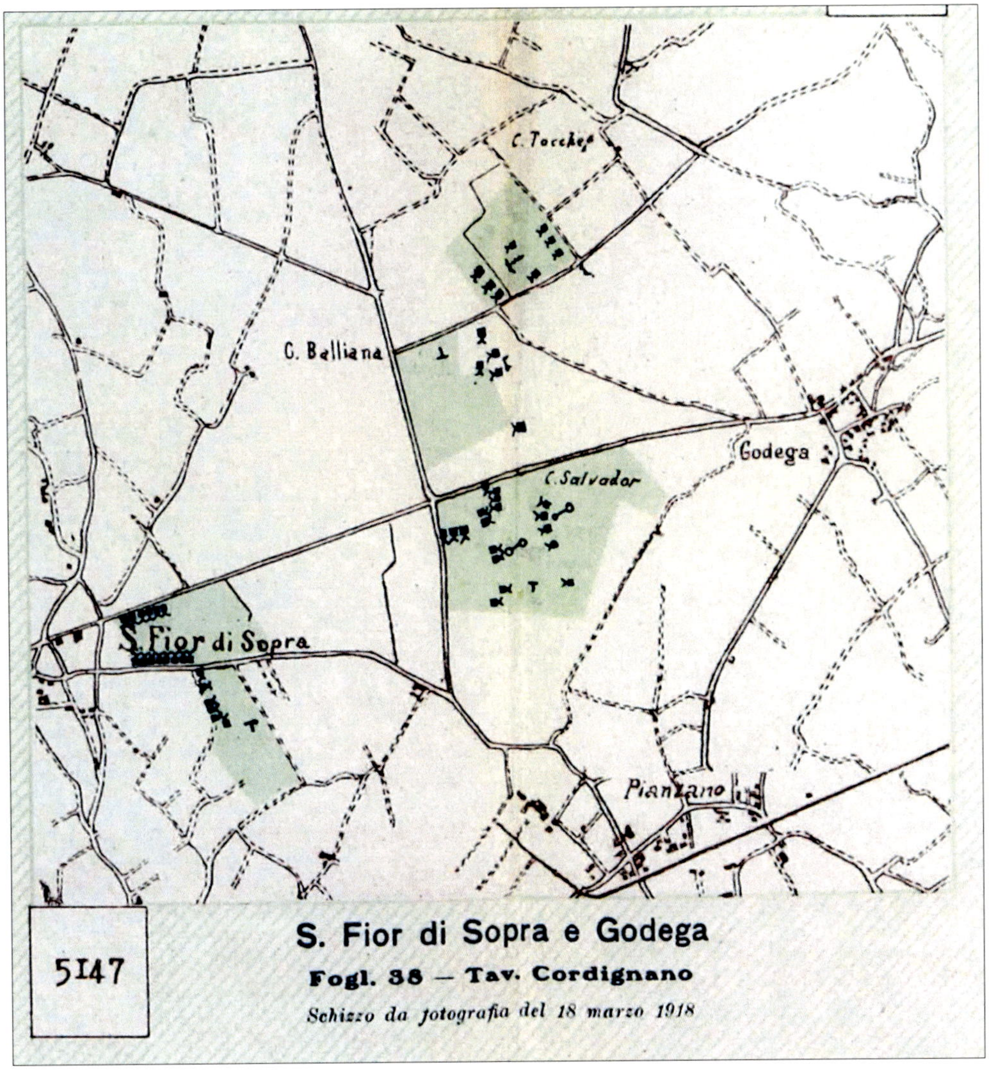

Right: Sketch of the objects on the airfields based on the reconnaissance photo of 18th March 1918. (Paolo Miana)

31, 39) moved to the airfield at San Fior.

The Allies also discovered the intense activity on this new airfield. Italian and British observers, during their reconnaissance sorties, counted 28 canvas hangars on 21st February and 20 on the 16th and 18th March.Most of these installations disappeared by the middle of April in parallel with the leaving of the German squadrons. Only three tents remained. When Flik 72/J and Flik 74/J arrived, the number of hangars increased again: enemy recce planes reported five on 9th May and nine at the end of the month – this figure did not change until September 1918.

Flik 74/J complained about circumstances month after month without improving or changing. In July and August, the unit had ten hangar tents, four of which were in such poor condition that the aircraft inside suffered seriously in the rain. Two of them became essentially useless by the end of August. The squadron reported 26 available fighters, despite the large number of aircraft, however, enemy air raids were quite rare on the base.

Above: Aerial photo of San Fior di Sopra airfield. (Paolo Miana)

Below: Winter street view in San Fior during the war (Giorgio Marinello)

Flik 68/J

Activation of Flik 68 as a fighter unit began sometime in December 1917 and the squadron was sent from Strasshof a.d. Nordbahn to the airfield of Godega di Sant'Urbano on 28th February 1918 with four pilots but without any planes or hangar tents. The commander of Flik 68/J was *Oblt.* Karl Patzelt and the pilots were,, at the end of March 1918,, the following: *Offstv.* Emanuel Stumpa, *Stfw.* Karl Benko, *Zgsf.* Johann Juhász, and *Zgsf.* Jakob Wesely. Instead of operational use, however, the unit was relocated to Aviano on 11th March, becoming the basis of the fighter school of the Austro-Hungarian 6th Army *(Jagdfliegerschule Aviano)* and was equipped with Albatros D.III (Oef) series 153, originally assigned to Flik 74/J. On 3rd April, *Stfw.* Josef Kočur, who was taking part in fighter training at the squadron, crashed during a practice flight due to wing failure of his plane. Albatros D.III (Oef) 153.43 was totally damaged, and the pilot lost his life in the accident.

Flik 68/J was transferred back to Godega on 14th April, and new pilots, planes and equipment were assigned to the squadron. *Fwbl.* Rudolf Blass was one of the newly arrived pilots and he recorded the events of the first two weeks in his diary:

"**12. IV.** *I flew to Aviano in the afternoon because I had been assigned to Flik 68/J, with the promise that I could go back to the* [Flieger] *park in two months. My Oberleutnant said he would do everything to get me back.*

15. IV. *I had gone to Pordenone by car the day before yesterday, it was raining all day yesterday, and I joined the squadron at Godega today because it had been relocated here from Aviano in the meantime.*

16. IV. *I flew with another guy in a "little Brandi"* [Hansa-Brandenburg B.I] *trainer plane to Casarsa for my mail in the afternoon, from here to Roveredo, then to Aviano to visit our fellows. This plane served for that sort of things, and it could be resorted without any special permission whenever in spare time. A "Berg" [landed] at the neighbouring fighter squadron in the evening. The pilot, Stfw. Ferenc Vally died.*

19. IV. *I reported myself to the commander, Oberleutnant Patzelt, on the 17th. It had been raining yesterday, but the weather was very nice today. There is temporarily no flying, as the squadron has just been established and our planes are not ready yet. The squadron has "D.III" fighter planes. In the morning, I was called to the airfield to choose a plane for myself, which was made ready by Tuesday. We are 10 NCO pilots, and there are three officer pilots. Two officer pilots and three NCO pilots, including me, were marked out for flying on Tuesday. My plane has the number 153-193, having a 200 hp "Austro-Daimler" engine flying with 200-220 km/h.*

Above: Portrait photo of Rudolf (Blass) Bakonyi around 1937–1938 (Bakonyi family archive)

21. IV. *It had been raining again yesterday, it had cleared up by today, and four of us went to Flik 4 at Belluno by car at 7 a.m. to pick up D.III planes and fly through the Tyrolean mountains. Unfortunately, we did not get the planes, so we returned by car. It would have been nice to fly over the ca. 2500-3000 m*[eter] *high mountains with a plane, although*

Above: Pilots of No. 28 Squadron in Italy (from left to right): Lt. N.H. Hamley, Capt. J. Mackereth, Capt. J.E. Hallonquist, Lt. L.P Wilson, Lt. G.F.M. Apps (66 Sqn), Lt. A. Latimer, Lt. S. Stanger, Lt. A.J. Yates, Lt. C.G. Constanduros, Lt. A.G. Cooper, Lt. P.G. Mulholland, Lt. T.F. Everett. Several of them took part in the first air combats of Flik 68/J. (Andy Kemp)

we could have probably flown through the same pass which we came. It was evening when we arrived back.

23. IV. *Six of us performed practice flying in formation. We flew to Pordenone and Aviano. The performers were three Oberleutnants, one Leutnant, one Offizierstellvertreter and me – Feldwebel. (Some further officers joined the squadron in the meantime.)*

I flew in somebody else's plane because mine was not readu. We had just taken off when I saw that the others landed one after another, so did I, of course. The chair had broken in one of the Oberleutnants' planes, which was why he landed. We did not know that and landed too!After the second take-off, in formation, we landed on the mentioned airfields. Arriving home, we flew a circle over the airfield, and we landed one by one – in the prescribed order. We received a compliment from the squadron commander for the nice formation flight.!

25. IV. Yesterday it had rained this afternoon cleared it up, and at four o'clock, seven of us – four officers and three non-commissioned officers – took off to carry out the task of the day before yesterday. We were returning over our airfield when I saw the two aircraft in front of me and slightly lower collide in the air. One of them was smashed to pieces in mid-air, and the other plane fell upside down. At the moment of the collision, one of the planes caught fire in and went down in flames. Both pilots died horribly.

"We landed immediately, of course, and in my nervousness I landed the plane so hard that the axis of the undercarriage twisted; it had been a long time since I flopped a plane. I came to know down on the ground that Leutnant Hans Kiebacher had hit Leutnant Egon Wagner's plane. The aircraft which caught fire entirely burnt on the ground."

In the accident, the fuselage of Oblt. Egon Wagner's Albatros D.III (Oef) 153.183 had broken, the aircraft and its pilot fell from an altitude of about 500 meters, smashing both on the ground. Due to the collision, the left wing of Lt. i.d.Res. Hans Kiebacher's 153.184 broke, and its fuel tank exploded about 50 meters above the ground, causing the terrible death of the pilot.

Following this tragic event, the squadron flew the first sorties on 1st May and the pilots encountered enemy planes: Stfw. Andreas Dombrowski had two air fights during the day. In one of the missions, the flight of Flik 68/J had an engagement with four "Sopwith" fighters, however, enemies avoided combat by steep diving.The next day brought another dogfight during an escort mission over the Monte Tomba. Austro-Hungarian

Above: Oblt. Karl Patzelt (fourth from the right) in front of an Albatros D.III (Oef) fighter with his comrades of Flik 29 on the Romanian front (Bernhard Tötschinger via Boris Ciglić)

reports mention three "Sopwith" scouts as opponents, which were chased away by the black-crossed fighters. *Lt. i.d.Res.* Kajetan Kosinski force-landed after the combat with his Albatros D.III (Oef) 153.176 due to engine failure; the pilot was injured, and his aircraft was damaged. The enemy planes probably were Camels of No. 28 Squadron, Capt. S. Stanger (B.6413) and Lt. C. McEwen (B.7383) had flown an extra patrol looking for two-seaters. At 10.30 a.m., they saw an enemy formation consisted of an Aviatik reconnaissance aircraft and 10 Albatros D.V fighters flying from Cismon in an easterly direction. Capt. Stanger attacked the Aviatik that immediately dived, leaving the British pilot who later re-attacked the two-seater, forcing it to dive again and crash near Valdobbiadene at 10.35 a.m. In the combat with the D.Vs, *Lt.* McEwen claimed a fighter downed in flames at 10.40. a.m. and the duel continued until Capt. Stanger ran out of ammunition. *Lt.* McEwen had only a few rounds left. The Camels broke off the combat then. Contrary to the British claiming the squadrons of the Austro-Hungarian 6th Army did not report any casualties in air combat that day in this area.

In the morning of 4th May 1918, four D.IIIs of the squadron flew a fighter sortie when they encountered a group of British Sopwith Camels. By the Austro-Hungarian description of the events, planes of Flik 68/J, under the command of squadron CO *Oblt.* Karl Patzelt, had a dogfight with six British Sopwiths over enemy territory. During the fierce air battle a group of another six enemy fighters joined the combat. *Stfw.* Dombrowski managed to shoot down an enemy plane in flames near Montebelluna, he was wounded in the cheek by a grazing shot.After the failure of his machine guns, he was forced to make an emergency landing at Piave di Soligo on the eastern bank of the Piave. Dombrowski's 153.195 was heavily damaged during landing. The air victory was confirmed by one of his comrades and an infantry report, and Dombrowski was awarded by Silver Bravery Medal 2nd Class. Other members of the Austro-Hungarian formation were not so lucky: the commander *Oblt.* Patzelt (five confirmed victories at Flik 29 and Flik 42/J) and *Kpl.* Franz Fritsch were missing in the fight,

Above: Stfw. Andreas Dombrowski (Christian Reiterer via Marcus Zelezny)

they probably lost their life. This was Fritsch's second mission, and according to his posthumous decoration proposal for Silver Bravery Medal 1st Class, he forced two enemy planes to make an emergency landing in enemy territory, confirmed by two survivor mates.

British pilots gave quite a different account of the battle. Four pilots of No. 66 Squadron took off at 8.55 a.m. for the Eastern Patrol: *Lt.* W.C. Hilborn (B.5220), *Lt.* V.S. Parker (B.7358), *Lt.* G.A. Birks (B.6424), *Lt.* G.F.M. Apps (B.2497). The patrol was flying over Vidor at 9.45 a.m. at an altitude of 14,500 feet when they observed 12 enemy scouts (mixed D.IIIs and D.Vs) at 16,000 ft. Then the combat went as follows: "*Coming towards them from the direction of Conegliano, patrol engaged the E.A. Lt. Apps shot one D.V down in flames and shot one D.III down. Lt. Birks shot one D.III down in flames, Lt. Parker shot one D.III down and Lt. Hilborn drove one D.III down out of control. The D.III shot down by Lt. Apps was observed to crash by Lt. McLeod of No. 28 Sqdn. The E.A. pilot got out of the wrecked machine and Lt. McLeod shot him. Lt. App's machine was shot through the engine, cowling, centre section, L.H. bottom plane, elevators, and one flying wire shot away.*"

Lt. G.D. McLeo (B.2316) of No. 28 Squadron carried out an escort mission with *Lt.* A. Latimer (B.6356), who had to return because his engine was malfunctioning. *Lt.* McLeod lost sight of the escorted Bristol F2b fighter over Motta di Livenza, and after that, "*Lt. McLeod climbed to 17000 ft and encountered 2 D.V's over the west end of the Montello at 9.50 a.m. He attacked the rear E.A. and fired a burst of 30 rounds at 40 yards range. The E.A. went down vertically, and was last seen at 12,000 ft still diving. Lt. McLeod then joined a Camel of 66 Squadron which was pursuing a yellow D.V and saw the E.A. crash about 3 km S.E. of Moriago about 10 a.m. Lt. McLeod circled round the crashed machine, saw the pilot emerge from the wreckage and shot him.*"

In addition, *serg.* Giovanni Nicelli joined the same battle when he took off for an engine test and realized that the artillery was firing, indicating the presence of enemy aircraft. Over the Montello, he found a group of enemy fighters and attacked them together with some Camels of No. 66 Squadron. He claimed two Albatros D.IIIs shot down within the Italian lines in cooperation with the British pilots, which were identified as the two lost members of Flik 68/J. The pilots saw *Kpl.* Fritsch, climbing over the wing to escape the flames and then jump down from an altitude of 100 meters. The unlucky pilot fell close to a large haystack, and he survived two hours after his fall.

At first, reading the British reports, their pilots seemed like they were taking part in a different combat, unless Austro-Hungarian documents are scrutinized. Because fighters of Flik 56/J had a dogfight with Sopwith Camels escorting a reconnaissance plane of Flik 4/D in the area of the Montello. Flik 42/J also carried out an escort mission having dog fights with Camels. Pilots of Flik 42/J claimed a Sopwith shot down east of Cornuda and another over the Montello. As British fighter units had no other combat even close to this territory (and neither the Italians), all black-crossed single-seaters must have taken part in the same battle, in which Austro-Hungarians claimed five enemy planes. They lost three of their own aircraft and pilots, while the British claimed four enemy fighters shot down (two of them shared with *serg.* Nicelli) and two out of control, and only one Camel was reported as damaged at the same time.

After the squadron lost his commander, *Oblt.* Georg Kenzian von Kenzianhausen of Flik 55J was appointed to the position. The bad weather did not allow any flying

Above: Dombrowski's damaged Albatros D.III (Oef) 153.195 after landing on 4th May 1918.

Known Missions Flown by Pilots of Flik 68/J in May 1918

Date	Flt Time	Task	Route or territory of deployment	Pilot	Aircraft
1.5.	1h 10'	fighter	territory of 6th Army (air fight)	Stfw. Dombrowsky	153.195
	50'	fighter	territory of 6th Army (air fight)	Stfw. Dombrowsky	153.195
				Stfw. Benko	153.194
2.5.	1h 15'	escort	territory of 6th Army	Stfw. Dombrowsky	153.195
				Stfw. Benko	153.194
3.5	1h 45'-2h	blockade defence	Colle Umberto–Pordenone–Conegliano	Stfw. Dombrowsky	153.195
				Kpl. Fritsch	153.210
	50'	fighter	territory of 6th Army	Stfw. Dombrowsky	153.195
4.5	1h 15'	fighter	territory of 6th Army (air fight 3 victories claimed)	Oblt. Patzelt (KIA)	153.182
				Stfw. Dombrowsky (WIA)	153.195
				Kpl. Fritsch (KIA)	153.210
13.5		escort		Offstv. Stumpa (KIA)	153.189
				Fwbl. Juhász	
				Fwbl. Wessely (INJ)	153.196

Above: Lt. W.C. Hilborn was a member of the victorious British patrol on 4th May 1918. (Andy Kemp)

activity for the squadrons of the Austro-Hungarian 6th Army in the next few days. On the 8th, however, the unit lost another fighter, without enemy action this time: Albatros D.III (Oef) 153.193 suffered serious damage during landing, its pilot, *Fwbl.* Jakob Wessely remained uninjured.

On 13th May, three Albatros D.III (Oef) fighters of the squadron carried out an escort mission of the reconnaissance aircraft of Flik 57/Rb. Two planes of the formation had to return after 1 ¾ hour flying due to fuel shortage. *Fwbl.* Juhász reached his own airfield at Godega and landed without fuel. *Fwbl.* Wessely had to make an emergency landing at Borgo-Saccon at Ballonkompagnie 25 with an empty fuel tank, the pilot was slightly injured, and his 153.196 was totally damaged. *Offstv.* Stumpa did not want to abandon his duty and remained alone protecting the two-seater but he was attacked by two Italian SVA 5 fighters over the Montello. In the course of the air fight, the engine of Stumpa's plane stopped in a turn due to lack of gasoline, and he was shot down in flames by his enemies. In consideration of this case, the air command of the Austro-Hungarian 6th Army ordered the squadron leaders and officer-observers to understand the fact that fighter pilots could only fly a maximum of two hours, and to provide for possible replacement of the escort to a long-range mission, avoiding the similar situations.

Some details of the fight and Stumpa's demise are unclear or confusing. It seems sure that the victorious Italian pilot was *ten.* Flavio Baracchini of 76a Squadriglia, the famous ace, who took off at 8.00 a.m. that day for a fighter mission probably with his Hanriot HD 6272. According to the War Diary of the Italian squadron, he attacked an enemy fighter over the Montello and shot it down near Arcade, firing 100 bullets. Following the action, he landed at 9.45 a.m. Other sources, like the daily report of 76a Squadriglia, stated that Baracchini forced another black-crossed fighter to land behind the Austro-Hungarian lines and this one is usually identified as Fwbl. Wessely and his Albatros D.III (Oef) 153.196. On the one side, there are reports of one Albatros being attacked by two SVA fighters while the other side reported that one Hanriot fought with one or rather two enemy single-seater(s) – so questions remained open.

After the duel, Baracchini visited Stumpa's wrecked plane and took some souvenirs from it: a wheel, a magneto and a piece of the propeller, for which action he was criticized by the commander of the Aquila Brigade. It is also interesting that in the photo of this wreckage the plane had wheel discs of the markings of Flik 57/Rb.

In less than three weeks, only two of them were operational, Flik 68/J lost ten planes and seven pilots (four died, one wounded, and two injured). It is no wonder that the squadron did not fly any sorties in the next month. The number of aircraft still dwindled. On 6th June, 153.205 was damaged in Campoformido by *Oblt. i.d.Res.* Karl von Westenholz who escaped without injury, and *Fwbl.* Gustav Döbriacher went missing during a formation flying practise in 153.108 on the same day. Three days later, 153.113 and 153.147 were also lost when they were damaged at 6.45 p.m. without any enemy action. On 10th June 1918, the unit handed

Above: Lt. G.D. McLeod (in the middle) shot down an enemy fighter on 4th May 1918. The other pilots in the group photo are Lt. H.B. Hudson (on the left) and Capt. N.C. Jones (on the right) (Andy Kemp)

Below: Serg. Nicelli stands in front of the wreck of Kpl. Franz Fritsch's Albatros D.III (Oef) 153.210 shot down on 4th May 1918. The next day the Italian ace also died. (Colin Overs via Boris Ciglić)

Above: Oblt. Georg Kenzian with Offstv. Josef Kiss of Flik 55/J on 12th January 1918 in front of the force-landed RE-8 A.4445 of No. 42 Squadron. Kenzian became the new and final commander of Flik 68/J after the death of Oblt. Patzelt. (Jiři Rajlich)

Right: Fwbl. Johann Juhász in a Hungarian magazine in 1919. He served all three fighter squadrons, Fliks 68/J, 72/J, and 74/J. (Aviatika)

Facing Page, Above: The wreck of Offstv. Emanuel Stumpa's Albatros D.III (Oef) 153.189. Note the squadron marking of Flik 57/Rb on the wheel disc. (Roberto Gentilli)

Facing Page, Below: Ten. Flavio Torello Baracchini of 76a Squadriglia with his brand new French-built Hanriot H.D. 515 emblazoned with the four aces. H.D. 6272, which he flew later, had the same personal marking. (Roberto Gentilli)

76
D
515

Known Missions Flown by Pilots of Flik 68/J in June 1918

Date	Flt Time	Task	Route or territory of deployment	Pilot	Aircraft
16.6	S: 20.10 L: 20.45	infantry fighter	Il Montello	Offstv. Pfundtner	222.08
				Zgsf. Pillwein	222.06
				+ 2 pilots	
19.6	S: 10.45 (40‘)	fighter	territory of 6th Army	Oblt. v. Kenzian	422.19
				Zgsf. Malowetzky	222.07
				Zgsf. Pillwein	222.06
	S: 18.15 L: 19.10	fighter	territory of 6th Army	Oblt. Kenzian	422.19
				Zgsf. Pillwein	222.06
20.6	S: 9.00 L: 10.25	bridge protection	Piave river	Oblt. Kenzian	422.19
				Offstv. Pfundtner	222.08
	S: 17.50 L: 18.30	bridge protection	Piave river	Oblt. Kenzian	422.19
				Offstv. Pfundtner	222.08
21.6	S: 11.00 L: 12.15	bridge protection	Piave river	Oblt. Kenzian	422.19
				Offstv. Pfundtner	222.08
	(1h)	bridge protection	Conegliano–Piave bridge	Zgsf. Pillwein	422.19
				+ ?	
22.6	S: 7.30 L: 7.45	defence	Conegliano–Piave bridge	Oblt. Kenzian	422.19
				Zgsf. Malowetzky	422.24
				Zgsf. Pillwein	222.04
	(45‘)	defence	Conegliano–Piave bridge	Zgsf. Pillwein	222.04
				+ ?	
23.6	S: 9.00 L: 10.10	blockade	territory of 6th Army	Oblt. Kenzian	422.19
				Zgsf. Malowetzky	422.24
				Zgsf. Pillwein	222.04
24.6	S: 6.25 L: 7.05	defence	territory of 6th Army	Zgsf. Malowetzky	422.24
				Zgsf. Pillwein	422.19
				Kpl. Kirchberger	222.04
	S: 7.40 L: 8.45	defence	territory of 6th Army	Zgsf. Pillwein	422.19
				Kpl. Kirchberger	222.04
	S: 17.00 L: 17.50	defence	territory of 6th Army	Zgsf. Pillwein	422.02
				+ 3 pilots	422.02
	S: 18.40 L: 19.25	defence	territory of 6th Army	Oblt. Kenzian	422.19
				Zgsf. Malowetzky	422.24
				Zgsf. Pillwein	422.02
				Kpl. Kirchberger	422.04

over its six remaining Albatros D.III (Oef) aircraft to Flik 30/J, which was being reinforced preparing for the Piave offensive. Flik 68/J was re-equipped with Phönix DII (series 222) and DIIa (series 422) fighters instead. The squadron commander, *Oblt.* Kenzian had a rather low opinion of the new aircraft some time later, and he

Known Missions Flown by Pilots of Flik 68/J in June 1918 (cont.)

Date	Duration	Task	Route or territory of deployment	Pilot	Aircraft
25.6	S: 7.30 L: 7.55	defence	territory of 6th Army	Zgsf. Pillwein	422.02
				+ 3 pilots	
	S: 8.00 L: 8.30	defence	territory of 6th Army	Zgsf. Pillwein	422.02
				+ 2 pilots	
27.6	S: 4.45 L: 5.45-6.10	escort	territory of 6th Army	Oblt. Kenzian	422.19
				Fwbl. Döbriacher	222.03
				Zgsf. Pillwein	422.02
				Kpl. Kirchberger	422.24

had a good reason for that: "*Flying them is almost suicide. The engine is installed in such a slapdash way, that its vibrations actually loosen it from its mountings, and it starts losing power. Consequently, the propeller starts to sway and gradually falls off. There is an easy solution to the problem, namely attaching the engine with bigger screws. But such screws are just not available at Fliegerpark! No matter how often I write, all in vain! My most experienced pilots will get shot down, and the replacements will probably be novices?!*"

Meanwhile, the last Austro-Hungarian offensive had begun. Flik 68/J and Flik 72/J were assigned to the Isonzo Army for the first day of the operation, since the attack against the Montello was intended only as a diversionary attack. The real cause of this shuffling of aviation units back and forth was that the number of squadrons was far too low for the offensive inthe beginning. In any case, Flik 68/J and Flik 72/J served under the 6th Army from 16th June again (according to the available orders of the battle, Flik 68/J was placed under the command of Austro-Hungarian XXIV Korps in the period of 1–20 July 1918.)

The squadron flew more than 50 sorties in the second half of June 1918, mostly defence missions for the bridges on the Piave, and only four times air fights

Known Missions Flown by Pilots of Flik 68/J in July 1918

Date	Duration	Task	Route or territory of deployment	Pilot	Aircraft
1.7	S: 10.25 L: 11.15	escort	territory of 6th Army	Fwbl. Döbriacher	422.24
				Zgsf. Pillwein	422.02
				Kpl. Kirchberger	422.03
2.7	S: 7.30 L: 7.55	escort	territory of 6th Army	Zgsf. Pillwein	422.02
				Kpl. Neftila	222.04
				Kpl. Thomas	222.06
4.7	S: 9.30 L: 10.00	defence	territory of 6th Army	Fwbl. Döbriacher	422.03
				Zgsf. Malowetzky	422.24
				Zgsf. Pillwein	422.02
				Kpl. Neftila	222.04
				Kpl. Thomas	222.06
5.7	S: 17.35 L: 18.10	defence	territory of 6th Army	Oblt. Kenzian	422.19
				Fwbl. Döbriacher	222.04
				Zgsf. Malowetzky	422.24
				Zgsf. Pillwein	422.02
				Kpl. Kirchberger	422.03
				Kpl. Thomas	222.06

Above: Albatros D.III (Oef) 153.108 of Flik 15/F on Levico airfield, spring 1918. This fighter was flown at Flik 68/J and was reported as missing during a formation flying practice on 6th June 1918. (Petr Aharon Tesař)

occurred with enemies. The pilots of the unit did not claim victories or suffer casualties during these combats, either. Accidents happened, though. On 19th June for example, four Phönix fighters led by the squadron commander *Oblt.* Kenzian took off to intercept enemy aircraft, but *Offstv.* Alfred Pfundtner had to return due to an engine failure. He made an emergency landing at the Pianzano airfield with his 222.08 ramming into Aviatik D.I 138.111, which was there waiting for take-off. *Offstv.* Pfundtner remained uninjured in the

Above, Below, & Facing Page: Albatros D.III (Oef) 153.113 after a crash-landing during a test flying at Wiener Neustadt in late 1917. The aircraft also served at Flik 68/J later and was damaged on 9th June 1918. (Martin O'Connor via Aaron Weaver)

accident, and both aircraft were reported to be seriously damaged, although 222.08 was flown again the next day by the same pilot (the plane was also slightly damaged by *Offstv.* Pfundtner in the evening of 18th June but it was successfully repaired by next day). His fortune did not last long: two days later, on the 21st, *Offstv.* Pfundtner crashed near S. Martino during his fourth mission, presumably due to a wing fracture, as the wings were lost already in spiralling down. Phönix D.II 222.08 was totally damaged, killing its pilot.

At least five further fighter planes of Flik 68/J were damaged in June without enemy action:- 153.205 was

Left: Oblt. i.d.Res. Karl von Westenholz (on the right) of Flik 66/D on the airfield at Feltre. The pilot served at Flik 68/J for a short time. Others in the photo (from left): Lt. i.d.Res. Ladislaus Kullmann and Stfwbl. Karl Urban.

slightly damaged by *Kpl.* Neftila on 25th June during landing after a practice flight; – 222.14 was totally damaged on 19th June in a landing after a mission; – 422.02 was slightly damaged in the hangar on 29th June; – 422.24 damaged during a practice flight in June; – 422.26 was also slightly damaged on 18th June during a training flight; on the 25th *Kpl.* Karl Thomas crashed with this plane when he was landing due to an engine failure, the pilot remained uninjured, the aircraft was totally damaged. By July 1918, several accidents had happened with Phönix D.II/D.IIa aircraft which were traced to engine bearer failure. Thus Phönix was required to supply reinforcement kits at no charge, although Phönix's factory was officially held blameless. The modifications were completed in July. The remaining fighters of Flik

Below: The remains of a Phönix fighter identified as D.II 222.04 of Flik 68/J (it seems more likely to be 222.08) (Christian Reiterer via Marcus Zelezny)

Right & Below: Pilot certificate of Kpl. Ladislaus Neftila, who served at both Flik 68/J and Flik 74/J (MMKM)

NEMZETKÖZI LÉGHAJÓS SZÖVETSÉG
MAGYARORSZÁG
294 sz.

Alulírott, a Nemzetközi Léghajós Szövetség által elismert magyarországi aéronautikai és aviatikai sporthatóság igazolja, hogy Neftila László úr szül. Sátoraljaujhely 1895-ben magyar honos miután az N. L. Sz. által előírt összes követelményeknek megfelelt, őt repülőgép VEZETŐNEK elismerte.

Budapest, 1918 május 6.

MAGYAR AERO-SZÖVETSÉG

elnök titkár

FÉDÉRATION AÉRONAUTIQUE INTERNATIONALE
HONGRIE
Nr. 294

Nous soussignés, pouvoir sportif reconnu par la Fédération Aéronautique Internationale pour l'Hongrie, certifions que Mr. Ladislaus Neftila né à Sátoraljaujhely le 1895 de nationalité hongrois ayant rempli toutes les conditions imposées par la F. A. I. a été breveté PILOTE aviateur

Budapest le 6 mai 1918

ASSOCIATION AÉRONAUTIQUE DE HONGRIE

Président Secrétaire

The Civil, Naval and Military Authorities including the Police, are respectfully requested to aid and assist the holder of this Certificate.

*

Gli agenti della forza pubblica e le autorità civili e militari sono pregati di voler dare aiuto ed assistenza al titolare del presente libretto.

*

Autoritățile civile și militare sunt rugate de a da portătorului acestui certificat scutul și sprijinul lor.

KÖZLEKEDÉSI MÚZEUM
537/okm

Neftila László
A tulajdonos névaláírása:
Unterschrift des Inhabers:

Sorties Flown by Pilots of Flik 68/J

Name	Arrived	1918			Left
		May	Jun	July	
Oblt. Karl Patzelt	26.1./?.2.1918	6	-	-	4.5.1918 †
Oblt. Georg v. Kenzian	5.5.1918	-	14	1	
Offzstv. Emanuel Stumpa	26.1./14.2.1918	8	-	-	13.5.1918 †
Offzstv. Alfred Pfundtner	?.6.1918	-	4	-	21.6.1918 †
Stfwbl. Andreas Dombrowsky	8.4.1918	6	-	-	21.5.1918
Stfwbl. Karl Benko	26.1.1918	2	-	-	21.5.1918
Fwbl. Jakob Wesely	26.1.1918	7	-	-	21.5.1918
Fwbl. Johann Juhász	26.1.1918	4	-	-	21.5.1918
Fwbl. Gustav Döbriacher	?.6.1918	-	4	2	26.7.1918
Zgsf. Johann Patleich	16.4.1918	4	-	-	?.5./21.6.1918
Zgsf. Ignaz Pillwein	?.6.1918	-	14	4	?.7.1918
Zgsf. Johann Malowetzky	?.6.1918	-	9	2	29.7.1918
Kpl. Franz Fritsch	3.3.1918	2	-	-	4.5.1918 †
Kpl. Josef Kunz	?.4.1918	1	-	-	?.5.1918
Kpl. Hans Kirchberger	5.6.1918	-	7	1	?.7.1918
Kpl. Ladislaus Neftila	?.6.1918	-	1	2	29.7.1918
Kpl. Karl Thomas	21.6.1918	-	-	3	26.7.1918
Altogether		40	53	15	

68/J were also sent to Fliegerpark 5 for this re-work by 10th July (422.19 and 422.24 on the 8th, 222.04, 222.06, 422.02, 422.03 on the 9th, while 222.07 remained at the squadron without engine). From 6th July, the unit was no longer operational, its pilots were transferred to other units, mostly to Flik 72/J and Flik 74/J, and the squadron was assigned neither new aircraft nor pilots until the end of the war. Their Phönix fighters were handed over to Flik 30/J after the reinforcement work.

Flying Activities of Flik 68/J, April–September 1918

Period	At Last Day of Period		Flying Days Avail.	Flown			Fights	Vics	Losses in Combat
	Aircraft	Pilot		Sorties	Flt Time	Dist. (km)			
1-30.4.1918	12	14	5	-	-	-	-	-	-
1-31.5.1918	5	14	23	36	61h	?	5	2	3
1-30.6.1918	4	7	26	52	38h 5’	?	4	-	-
1-31.7.1918	4	8	28	15	9h 10’	?	-	-	-
1-31.8.1918	-	-	29	-	-	-	-	-	-
1-30.9.1918	6	-	28	-	-	-	-	-	-

Flown sorties according to the half-monthly reports: 1-15.5.1918: 28 (61h), 15–30.6.1918: 79 (89h 20’), 1–s15.7.1918: 14 (8h 20’)

Flik 72/J

The squadron, like Flik 74/J, was established in January 1918 at the airfield of Straßhof a.d. Nordbahn, north of Vienna. However, as with all Fliks set up "late", becoming operational needed quite a long time because of the material and human misery of k.u.k Luftfahrtruppe at that time. In these cases, it took about two months until sufficient personnel and equipment were compiled. Flik 72/J carried out its first action in the second half of April 1918 under the command of the Austro-Hungarian 6th Army, though figures of the army's half-monthly report show no one in this period. The squadron commander was *Oblt. i.d.Res.* Josef Hoffmann ritter von Ostenhof, who held this position throughout the existence of the unit.

Among the sporadic sorties in May, the 20th became a heavy day for the fighter squadrons located at San Fior and Pianzano. Nine aircraft of the Austro-Hungarian XXIV Corps were ordered for free hunting that day. At 11.45 a.m., six planes of Flik 56/J (three of them had to return earlier due to engine trouble), *Oblt. i.d.Res.* Josef Hoffmann (115.25) of Flik 72/J, *Oblt.* Franz Cserich (138.119) and *Lt. i.d.Res.* Ernst Piech (138.63) of Flik 74/J took off to accomplish the task that eventually brought many fights.

At 10.30 a.m., four Sopwith Camels of No. 66 Squadron took off for the Western Patrol and chanced upon the Austro-Hungarian fighters: *Capt.* Barker (B.6313), *Lt.* G.A. Birks (B.6424), *Lt.* W.C. Hilborn (B.5220), *Lt.* G.F.M. Apps (B.5190). Capt. Barker reported the combat as follows: *"While patrolling just N. of M. BALDO at about 15,500 ft. patrol observed 7 E.A. scouts at 16,500 ft at 12 noon N. of CISMON. E.A. were proceeding East and patrol followed them climbing. Patrol were almost up to the level of 3 of the E.A. just N. of VALDOBBIADENE. Lt. Hilborn fired up to at one E.A. who went down in a spin. Pilot followed E.A. firing several bursts and at 7,000 ft. E.A. went down completely out of control and crashed in a field at 38.H.1585. This was observed by Capt. Barker and Lt. Apps. Lt. Birks attacked another E.A. from just below. E.A. continued to climb and then suddenly did two complete rolls. Lt. Birks fired a long burst at E.A. when he was coming out of the second roll and E.A. immediately went down out of control, crashing in a field at 38.H.5090. This was seen by Capt. Barker. Lt. Apps fired at another E.A. who was on Lt. Hilborn's tail. E.A. went down but pilot lost sight of him under his wings. Capt. Barker, who was below the patrol, tried to reach the remaining four E.A. but only managed to fire stalling shots at one E.A. who went down in a spin. Pilot was unable to follow this E.A. as another E.A. was taking up a position on his tail. Capt. Barker and Lt. Apps followed the remaining E.A. beyond SACILE but they refused combat."*

Above: Oblt. i.d.Res. Josef Hoffmann-Ostenhof, commander of Flik 72/J. (Christian Reiterer via Marcus Zelezny)

According to Austro-Hungarian documents, three pilots had a dogfight with the Camels. *Oblt.* Cserich had a duel with a Sopwith and *Kpl.* Zavodsky (138.99) of Flik 56/J with two – the last-mentioned plane was hit several by the British. *Oblt. i.d.Res.* Hoffmann's Aviatik D.I suffered the most serious damages during the combat, the right lower wing of his aircraft broke, and the squadron commander had to make strong efforts to reach his own airfield. However, none of the Aviatik fighters crashed, and this fact contradicts the British claims – like in many cases.

On 29th May, the unit reported five combat-ready fighters (Aviatik D.Is with 200 hp engine) and seven pilots (including three officers), one further assembled

Above: Capt. W.G. Barker with his Sopwith Camel B.6313. (Aaron Weaver)

aircraft without machine guns and two under repair at the company, waiting for the arrival of the missing components (undercarriage and wings).

The month of June proved much more eventful for Flik 72/J. On 2nd June, five aircraft of the squadron spotted four enemy planes during a mission, and the Austro-Hungarian pilots followed them, but were unable to catch them up. On the way back, Aviatik D.I 115.25 had an engine failure and its pilot, *Fwbl.* Ernst Heinz was compelled to make a force-landing, where the plane was totally damaged, the pilot injured and was carried to the hospital.

Some days later, on 6th June, four pilots of the unit carried out another free hunting and fought with two enemy aircraft over the Montello. One of the planes, Aviatik D.I 138.36 was damaged during force-landing, but the reason for that and also the identity of its pilot, who was injured, is unknown.

Beginning with the Austro-Hungarian offensive on 15th June at the Piave, the number of missions increased significantly. On the second day, the pilots of Flik 72/J took off three times, and they flew seventeen sorties to protect the bridges on the Piave. Planes which rose into the air at 11.15 a.m., had many combats with enemies. *Fwbl.* Wessely fought at first with two reconnaissance planes, then *Zgsf.* Patleich had an engagement with a "Sopwith" and *Fwbl.* Wessely with two ones. As British squadrons had no fights on that day, the foes had to be Italians, but their identification is not possible from the available documents.

On 19th June, four NCO pilots led by *Oblt. i.d.Res.* Bistritschan fulfilled a mission to protect the crossing point on the Piave at Mercadelli. Over Spresiano, they met an enemy reconnaissance aircraft escorted by three Hanriot fighters and claimed to bundle the Italians off.

The next day brought the highest number of air combats during the Austro-Hungarian offensive in June at the Piave and certainly, the squadrons of San Fior di Sopra also took part in them. The rain in the morning did not obstacle the activities of Italian planes. Besides reconnaissance and direction of artillery, bombers also departed their base to attack bridges on the Piave despite the low clouds. Seven Italian Capronis of IV Gruppo and six of XIV Gruppo raided targets at Falzé,

Above: Fwbl. Ernst Heinz's crash-landed Aviatik D.I (Lo) 115.25 on 2nd June 1918.

dropping 70 bombs (3890 kg). Four fighter squadrons of the Austro-Hungarian 6th Army flew many defensive and escort missions for the Piave bridges that day. *Fwbl.* Josef Peiger (238.54), *Fwbl.* Johann Juhász (115.10) and *Zgsf.* Johann Paleich (184.01) took off at 9.50 a.m. for a fighter sortie in the area of Piave–Montello. They met a flight of bombers and fought with six Capronis and four escort fighters identified as Sopwiths. *Zgsf.* Patleich had another combat with a Savoia reconnaissance aircraft at an altitude of 2500 meters, but the enemy went down in a dive and disappeared in the clouds. Five planes of Flik 72/J flew a free hunting mission in the evening hours, too, but they did not find any foes this time.

On 21st June, a flight of five planes took off twice to defend a bridge on the Piave at the Montello. First, between 11.00 and 12.00 a.m., they saw two enemy fighters and a reconnaissance plane, but they did not come into contact with each other. At 7.00 p.m., the same flight went up with the same task, this mission, however, turned into a serious dogfight. *Oblt. i.d.Res.* Hoffmann, *Fwbl.* Juhász and *Zgsf.* Patleich had an air fight with a "Sopwith" over the Montello, that was driven away. Then *Oblt. i.d.Res.* Hoffmann had another combat with a Nieuport sending it to a spin and claimed a probable victory, but Hoffmann's plane was also damaged. *Fwbl.* Wesely was reported as missing in action, his aircraft (115.31) was found later in the northern part of Planquadrat 1697, at Casa Campana. According to the half-monthly army report, the pilot was shot down in a combat with a Nieuport. His companions landed at 8.15 p.m. on the airfield of San Fior.

This evening *ten.* Silvio Scaroni flew an offensive patrol with *ten.* Alessandro Buzio, on the other side from the Montello to Susegana, and spotted an Austro-Hungarian flight at about 7.30 p.m., returning to their lines. As the enemy fighters crossed the lines at 7.55 p.m., they attacked one of them and shot it down. Scaroni remembered as follows in his memoirs: "*It will not take me long to find out because our summed speed burns very quickly the distance that separates us. In a flash, I find myself within range of their machine guns, but instead of shooting at me, the group divides, passing me partly on the left and partly on the right side.*

I have the impression for a moment that they have made a very skilful manoeuvre to surround me, but as soon as the last of the group on the left passes under my wing and slips behind the tail, without waiting for their further move, I turn abruptly and dive on him at full throttle, and even before he begins any of the movement I give him, almost abeam, a good burst.

Above: Lt. i.d.Res. Friedrich Bistritschan, observer of Flik 14, later pilot at Flik 72/J and Flik 74/J in a rank of Oblt. i.d.Res. (Jiři Rajlich)

The enemy aircraft continues its straight flight for a short while, then it goes down and disappears into the clouds below in a moment.

The crackle of my machine gun forced the other five to engage me, of course, but before they even decide what to do, I have already dived into the clouds, too, following the fighter that I shot.

When I get out of that dense, humid fog, I have the impression of falling into a dream world: below me, the plain stretches out confused, flooded with mist. I cannot find the orientation for a few moments because the landscape below shows no relief; the greyness of the fog flattens the immense expanse of land that I glimpse below me.

The rumble of my engine is almost overwhelmed by the noise of the propeller and the steel cables, which scream like the rigging of a ship seized by a storm. I came out of the clouds with the aircraft descending vertically, so I have to manoeuvre with all the prudence not to risk seeing my wings leave.

There are no planes in sight; I already despair to see my opponent again, when there he is, a thousand meters below me, trying to cross the Piave.

The steel cables of my plane start whistling again, and in a few moments, I reach the right firing distance. But at that moment, the enemy pilot notices my presence and begin a turn to avoid my burst: it is now too late for him... he is already hit directly...

The enemy aircraft continues on its own the turn that was initially enforced by the pilot until it overturns completely. Still descending with the wheels pointing upwards, it crashes miserably south of Mandre, two kilometers beyond the river, exactly on the side of the main road."

Although Scaroni mentioned in his memoirs that he had lost his companion in the clouds, his logbook reports the participation of another Hanriot, probably the one of Buzio. The victory was witnessed and confirmed by observers and also *ten.* Bacula of 91ª Squadriglia, who arrived in the final part of the combat.

By 22nd June, the number of combat-ready aircraft at Flik 72/J decreased so much that no more formation sortie was possible in the month. Only a trainer aircraft with a 165 hp engine was in operational condition, and another Aviatik D.I with a 200 hp engine was available on some days.

On 25th June, three-engined bombers of Italian IV and XIV Gruppo were ordered to attack munition depots at Visnadello, Orsago and ones along the line of Conegliano–Vittorio. About 8.15 a.m. (there are many variants of time in different reports), a flight of three Capronis appeared over the airfield of Flik 38/D at San Vendemiano and dropped bombs. Two 50 kg bombs fell to the airfield, one of them exploded close to a hangar tent stored Hansa-Brandenburg C.I 369.118 inside – both the tent and the aircraft were totally damaged, besides two men were also killed. The other bomb did not explode, some further fell to and around the officer's house.

Several planes of the Austro-Hungarian fighter units departed their airfields against the Capronis (four of Flik 56/J, seven of Flik 68/J and one of Flik 72/J), but only a few managed to reach and get in touch with enemies. The army daily report mentioned only one combat between five enemy and three own aircraft at 8.30 a.m. over Conegliano, which was probably the fight between Italian Hanriots of 76ª Squadriglia and pilots from Flik 42/J who also took part in chasing for the intruders, with Zgsf. Walter Babel of Flik 56/J joining that duel.

Kpl. Rudolf Dohnal of Flik 72/J took off at 8.00 a.m., and found the attackers. He had a short, pointless combat with a Caproni, so got back without success.

The month of July was much less intensive for the squadron over the front. That does not mean, however, that the period was uneventful. The pilots needed

Known Missions Flown by Pilots of Flik 72/J in May–June 1918

Date	Flt Time	Task	Route or territory of deployment	Pilots	Aircraft
20.5	S: 11.45	free hunting		Oblt. Hoffmann	115.25
20.6		fighter	Sacile–Oderzo	Fwbl. Heinz (INJ)	115.25
				+ 4 pilots	
16.6	S: 7.00		to the west egde of the Montello	Oblt. Hoffmann	
				+ 6 pilots	
	S: 11.15	bridge protection		Fwbl. Wesely	
				Zgsf. Patleich	
	S: 12.30			Oblt. Hoffmann	
				Kpl. Dohnal	
19.6	S: L: 16.05	defence	over Piave, crossing point at Mercadelli	Oblt. Bistritschan	
				+ 4 NCO pilots	
20.6	S: 9.50 L: 10.30	fighter bridge protection	Piave–Montello (air fight)	Fwbl. Juhász	115.10
				Fwbl. Peiger	238.54
				Zgsf. Patleich	184.01
	S: 17.55 L: 18.30	free hunting	Piave–Montello	Oblt. Hoffmann	138.61
				Fwbl. Juhász	115.10
				Fwbl. Wesely	115.31
				Zgsf. Patleich	184.01
				Kpl. Dohnal	238.54
21.6	S: 11.00 L: 12.00	bridge protection	Piave–Montello–Valpago	Oblt. Hoffmann	138.61
				Fwbl. Juhász	115.10
				Fwbl. Wesely	115.31
				Zgsf. Patleich	184.01
				Kpl. Dohnal	238.54
	S: 19.00 L: 20.15	bridge protection	Piave–Montello (air fight)	Oblt. Hoffmann	138.61
				Fwbl. Juhász	115.10
				Fwbl. Wesely (KIA)	115.31
				Zgsf. Patleich	184.01
				Kpl. Dohnal	238.54
23.6	S: 7.30 L: 8.00	bridge protection	Piave–Montello	Fwbl. Peiger	184.01
				Kpl. Dohnal	238.54
24.6	S: 17.15 L: 18.05	defence	along the Piave to the own troops	Kpl. Dohnal	238.54
25.6	S: 8.00 L: 8.45	defence	(air fight)	Kpl. Dohnal	238.54

training, and several accidents happened during these flying. At least five planes were written off in July: 115.03, 115.12 and 238.05 were damaged during landing, 138.16 crashed due to unknown reasons, like details of damaging of 238.22 are also unknown (series 238 planes were used as training aircraft, others are unsure). According to the

Above: Personnel of Flik 72/J on San Fior airfield, summer 1918. Sitting in the middle is the squadron commander, Oblt. i.d.Res. Josef Hoffmann, on his left, is Fwbl. Josef Peiger. The aircraft in the background is probably a series 138 „Berg". (Reinhard Masajdek via Petr Aharon Tesař)

army daily report of 23rd July, pilot(s) of Flik 72/J flew escort mission(s) that day, and one aircraft had a combat with three enemy planes. The wings and the engine of the black-crossed fighter were shot up, the identity of the pilot and the "Berg" are unclear.

The most significant event for the unit was still to come. Italian bombers appeared on 29th July, and the target was the airfield of Flik 101/G at Pordenone this time. At 7.00 a.m., eleven Capronis escorted by seventeen fighters were reported (other documents mentioned thirteen Capronis and about twenty escorting fighters). Almost all operational fighter squadrons of the Austro-Hungarian 6th Army started against attackers, and at least three duels evolved. In the first

Right: Fwbl. Josef Peiger of Flik 29 on the Romanian front on 10th August 1917. He was later transferred to Flik 68/J, then to Flik 72/J and crashed with Aviatik D.I 138.98 on 29th July 1918. (Árpád Kajon)

Above: Line of aircraft on San Fior airfield, four fighters of Flik 72/J and one with unit markings of Flik 74/J on the right. The first on the left is Aviatik D.I 238.39 „4", and the second is a series 115 plane. (Reinhard Masajdek via Petr Aharon Tesař)

Below: Close-up of the gun installation of Aviatik D.I (Lo) 115.13, which served at Flik 72/J (Martin O'Connor via Aaron Weaver)

Known Missions Flown by Pilots of Flik 72/J in July 1918

Date	Duration	Task	Route or territory of deployment	Pilot	Aircraft
2.7	S: 7.30	escort (Flik 59)	territory of II. Korps	Oblt. Bistritschan	138.62
	L: 8.20			Fwbl. Peiger	115.35
13.7	S: 6.25 L: 7.45	offensive sweep		Oblt. Bistritschan	138.62
				Oblt. Dehne	138.61
15.7	S: 7.25 L: 8.15	defence		Oblt. Bistritschan	138.62
				Fwbl. Peiger	115.10
				Kpl. Dohnal	115.54
16.7	S: 9.20 L: 10.45	escort (Flik 52)		Kpl. Dohnal	115.15
29.7	S: 17.35 L: 18.10	defence		Oblt. Dehne	138.62
				Fwbl. Peiger	138.98

one, at 7.10 a.m., three defenders fought with a part of the enemy formation, where an own aircraft was shot down. Since known casualties were only from Flik 72/J in this area and the unit flew defensive sorties with Flik 74/J on Italian Capronis, the event probably covers this engagement. During the combat *Oblt.* August Dehne in Aviatik D.I 138.62 and *Fwbl.* Josef Peiger in Aviatik D.I 138.98 crashed at Vignovo; according to the official Austro-Hungarian version supposedly due to breaking their wings. Both pilots died, Dehne's plane was totally damaged, and Peiger's one burned out. Many details of this fight are uncleared, so identification of enemies

Above: Pilots of 76ª Squadriglia (from right to left): Ten. Alessandro Buzio, Ten. Michele Borea D'Olmo (81ª Squadriglia), Ten. Giuseppe Retinò, Ten. Silvio Scaroni, Barbanti medical officer, Cap. Giulio Lega. Scaroni and Buzio were the victorious pilots of the air combat on 21st June 1918. (Roberto Gentilli)

Above: Ten. Bacula of 91ª Squadriglia, the witness of Scaroni's air victory on 21st June 1918. (Roberto Gentilli)

strikes a snag. It is not known either how many planes of Flik 72/J and Flik 74/J took part in the action. *Oblt.* Arpad Pindter von Pindtershofen had an encounter with Caproni and Hanriot planes in the area of Pordenone and Sacile, and the whole sortie took 25 minutes.

Among Italian escort fighters of 82ª Squadriglia

Sorties Flown by Pilots of Flik 72/J, April–July 1918

Name	Arrived	1918				Left
		Apr	May	Jun	July	
Oblt. i.d.Res. Dr. Josef Hoffmann Ritt. v. Ostenhof	4.2.1918	2	7	13	-	
Oblt. i.d.Res. Fritz Bistritschan	29.3.1918	1	2	10	6	22.8.1918
Oblt. i.d.Res. Ferdinand Strommer	21.5.1918	-	1	1	-	3.8.1918
Oblt. i.d.Res. August Dehne	15.6.1918	-	-	-	1	29.7.1918 †
Stfw. Johann Juhász	21.5.1918	-	2	13	-	2.8.1918
Stfw. Andreas Dombrowsky	21.5.1918	-	-	-	-	?.5/6.1918
Fwbl. Josef Peiger	21.5.1918	-	1	6	6	29.7.1918 †
Fwbl. Jakob Wesely	21.5.1918	-	3	15	-	21.6.1918 †
Fwbl. Ernst Heinz	30.5.1918	-	-	1	-	16.8.1918
Fwbl. Johann Patleich	21.5.1918	-	-	13	1	2.8.1918
Kpl. Rudolf Dohnal	21.5.1918	-	-	8	4	2.8.1918
Altogether		3	16	80	18	

Above: Aviatik D.I 238.39 fighter trainer of Flik 72/J. The unit and personal markings are clearly visible. (Colin Overs via Boris Ciglić)

claimed victories the following pilots: *cap.* Carlo Tosi at Pordenone (unconfirmed), *ten.* Renato Rossetti in outskirts Mansué (unconfirmed), *ten.* Aldo Comandone north of Lago di Santa Croce. British pilots of No. 45 Squadron also claimed victories in the area that day: *Capt.* N.C. Jones an enemy aircraft at Brugnera and *Lt.* A.J. Haines two Phönix fighters at Prata and Pordenone, however at 9.00 a.m. – this was probably another combat with four scouts and a two-seater, and did not connect to the Italian raid.

Known missions flown by pilots of Flik 72/J in July 1918 are shown in the table.:

After this mission, especially due to the shortages of the necessary personnel and equipment, Flik 72/J de facto disbanded on 2nd August 1918.

Flying Activities of Flik 72/J, April–September 1918

Period	At Last Day of Period		Flying Days Avail.	Flown			Fights	Vics	Losses in Combat
	Aircraft	Pilot		Sorties	Flt Time	Dist. (km)			
1-30. 4. 1918	9	4	2	3	4h 25'	640	-	-	-
1-31. 5. 1918	12	8	13	16	19h 10'	1940	3	-	-
1-30. 6. 1918	4	7	11	80	74h 20'	8570	18	-	1
1-31. 7. 1918	4	8	19	18	9h 5'	?	3	-	3
1-31. 8. 1918	-	1	-	-	-	-	-	-	-
1-30.9.1918.	-	1	-	-	-	-	-	-	-

Flown sorties according to the half-monthly reports: 1–15.5.1918: 2 (3h), 16–31.5.1918: 13 (7h 20'), 1-14.6.1918: 15 (21h), 15–30.6.1918: 62 (54h 15'), 1–15.7.1918: 8 (7h 50'), 16–31.7.1918: 6 (5h 55')

Above: The smashed fuselage of Oblt. August Dehne's Aviatik D.I 138.62 on a road near Vigonovo on 29th July 1918 after Italian Hanriots of 82a Squadriglia. (Reinhard Masajdek via Petr Aharon Tesař)

Below: Lineup of Phönix fighters of Flik 30/J. Phönix D.IIs of Flik 68/J were assigned to this unit after their reinforcement work. (Martin O'Connor via Aaron Weaver)

Flik 74/J

This unit was also established in January 1918 at the airfield of Straßhof a.d. Nordbahn accomplished its first sorties in the second half of Mai 1918, directly under the command of Austro-Hungarian 6th Army. This subordination remained unchanged until the end of the war except for a short period: the available orders of the battle shows that Flik 72/J and 74/J were placed under the command of the Austro-Hungarian II Corps, Flik 72/J in the period of 1-20 July 1918, Flik 74/J already from 12th June until 20th July 1918. The squadron commander was *Oblt.* Franz Cserich from the beginning, an experienced observer and pilot who served earlier at Flik 15 and Flik 24.

The first sorties in May were uneventful, the first known combat of the unit evolved on 6th June, when a patrol of the squadron flew a mission over the territory of the Montello and *Stfw.* Benko attacked two enemy aircraft two times (other report mentions a SPAD fighter as foe) but these actions remained unsucceessful. *Oblt.* Pindter had also a dogfight with a Sopwith and he had to force landed on the airfield of Flik 32 after a wing fracture. His Aviatik D.I 138.36 was totally damaged during landing, however, Pindter escaped from the plane uninjured.

Pilots of Flik of Flik 74/J carried out many fighter and escort sorties in the following days, too, especially from the beginning of the Austro-Hungarian offensive at the Piave. Already the first day brought ten combat missions with the first claim and a casualty, too. The

Above: Oblt. Franz Cserich, commander of Flik 74/J.

Known Missions Flown by Pilots of Flik 74/J in May 1918

Date	Flt Time	Distance	Task	Route or territory of deployment	Pilot	Aircraft
20.5	1h 15'	120 km	escort fighter	Piave–Montello	Oblt. Cserich	138.119
					Lt. Piech	138.63
21.5	1h 15'	120 km	escort	Piave–Montello	Lt. Piech	138.63
	1h 30'	145 km	fighter	Piave–Mascrada	Lt. Piech	138.63
22.5	1h 30'	145 km		Treviso–Volpago–Spresiano	Lt. Piech	138.63
	1h 20'	120 km	escort	Treviso–Volpago–Spresiano	Stfw. Benko	138.44
23.5	40'	90 km	defence	Piave	Lt. Piech	138.63
	1h 30-40'	140-160 km	fighter	Montello–Cormada	Lt. Piech	138.63
					Stfw. Benko	138.44
24.5	35-45'	70-95 km	fighter	Piave–Montello	Lt. Piech	138.63
					Stfw. Benko	138.44

Above & Below: Stfw. Karl Benko in his Aviatik D.I 138.44. Note the lack of personal marking on the plane and the rudder with streaked pattern originated maybe from another aircraft. (Reinhard Masajdek via Petr Aharon Tesař)

Above: Lt. i.d.Res. Ernst Piech, who died on 15th June 1918 as an officer pilot of Flik 74/J.

victorious pilot was *Stfw.* Benko, but the circumstances of the fight are lesser-known. Benko flew in the area of Nervesa–Montello, alongside the Piave in his Aviatik D.I 138.44, when he observed an aircraft over the Montello flying in the western direction. He followed it from a higher position, and as he closed the plane, which he identified as an SVA single-seater (Benko's decoration proposal described the Italian plane as a SPAD). Benko did not hesitate and attacked the Italian fighter, managing to hit his foe that crashed at Vidor. Available documents do not enable to answer the questions like who was the enemy pilot, the exact time of the air combat and whether *Stfw.* Benko flew alone in this combat. At any rate, that was the first confirmed air victory of the squadron and Benko's fourth one.

On the same day, Flik 74/J suffered a serious casualty when *Lt. i.d.Res.* Ernst Piech crashed with Aviatik D.I 138.63 shortly after take-off, and the pilot died in the accident.

The squadron took part in attacking Italian troops, too. On 19th June, four planes strafed an Italian battalion marching from the direction of Treviso towards Nervesa from an altitude of 100 meters about 1 km to Nervesa. Pilots also dropped bombs on the columns, and Italian soldiers cut and ran.

Two days later, *Stfw.* Benko took off in Aviatik D.I 138.44 with one of his companions in the evening hours and flew in the area of Cornuda, where he sighted a "Sopwith" biplane. The enemy aircraft fled, and a pursuit began that led from Montebelluna to Trevignano. At about 500 meters height, Benko could achieve a decisive hit, the enemy biplane was falling in a spin, but its pilot managed to make a force-landing on the airfield of Trevignano. This aerial victory could not be confirmed for Benko, since there was no witness.

Things happened not always in the planned way due to different circumstances like weather, technical failures, etc. It was the case at Flik 74/J, especially on the 22nd. Problems already occurred during the mission in the morning hours when the participants were unable to carry out an organized formation flight. A similar sortie was in the evening hours when five aircraft took off at 7.35 p.m., but *Kpl.* Soviansky and *Kpl.* Staatz could not find their companions and returned after 10 minutes, while *Stfw.* Benko was forced to return by an engine failure spending 20 minutes in the air. So, *Kpl.* Wissinger was the only one who flew with his flight commander over the Montello.

On 23rd June, five aircraft departed the airfield to accomplish a defence mission for the bridges on the Piave. Their route was uneventful, only a "Sopwith" was detected, seemingly protecting a balloon. *Kpl.* August Staatz, however, failed to return. He landed on the racecourse of Treviso, which was also a British airfield. The pilot became a prisoner of war uninjured, his Aviatik D.I 138.43 was slightly damaged and was later exhibited.

Although reports did not mention any fighter of Flik 74/J as a participant in the defence mission against the Italian Caproni bombers and their fighter escort on 25th June (or any other that day), *Kpl.* Adolf Wissinger fought with enemies in the area of Conegliano flying with Aviatik D.I 38.42 and claimed a SPAD shot down. It is not clear if Wissinger was the only one from the squadron in this combat, and identification of opponents is not possible, either, because of lacking further details.

With July came rather losses than successes. On the 12th shortly after take-off, the carburetor of *Kpl.* Rudolf Patzelt's Aviatik D.I 38.65 caught fire due to the poor quality of fuel. Patzelt crashed on San Fior airfield but had a fortune and suffered only some slight burn and leg injury. He was taken to a field hospital the same day.

Above & Below: Kpl. Karl Thomas' Aviatik D.I 138.43 after landing on the Treviso racecourse with only minor damage on 23rd June 1918. (Rinaldo D'Ami via Paolo Varriale)

Patzelt was entirely new at the unit since he arrived on 29th June and had not flown over the enemy territory.

The next day also brought a fierce battle for the units of San Fior di Sopra, but the details are quite confused. The daily report of the Austro-Hungarian 6th Army gives an account of the events: Flik 72/J and 74/J carried out fighter missions to protect a reconnaissance aircraft of Flik 59/D, and they had a scoreless fight with six Hanriot fighters over San Fior. Since there is no data on another dogfight, that took place here that

Known Missions Flown by Pilots of Flik 74/J in June 1918

Date	Flt Time	Distance	Task	Route or territory of deployment	Pilot	Aircraft
1.6	30'	35 km	fighter	Montello	Lt. Piech	138.63
	1h 6'	40 km	fighter	Montello	Stfw. Benko	138.44
3.6	1h 15'	30-40 km	escort	Cornuda	Oblt. Pindter	138.63
					Stfw. Benko	138.44
					+ 2 pilots	
6.6	40'-1h S: 11.20 L: 12.25	35-40 km	fighter	Montello (air fight)	Oblt. Cserich	138.119
					Oblt. Pindter	138.36
					Lt. Piech	138.63
					Stfw. Benko	138.44
					Kpl. Soviansky	138.43
8.6	45'	35 km	fighter	Montello	Oblt. Pindter	138.36(?)
9.6 (10.6?)	40' S: 6.50 L: 7.35	30 km	fighter	Montello	Oblt. Cserich	138.11?
					Oblt. Pindter	138.43
					Lt. Piech	138.63
15.6	S: 4.30 L: 5.40	30-45 km	fighter	territory of II Corps (air fight, 1 confirmed victory)	Oblt. Cserich	138.119
					Oblt. Pindter	138.36
					Lt. Piech	138.63
					Stfw. Benko	138.44
					Kpl. Soviansky	38.30
					Kpl. Wesely	38.41
	S: 10.30		fighter		Oblt. Cserich	138.119
					Oblt. Pindter×	138.43
					Lt. Piech (KIAcc)	138.63
					Kpl. Wesely	38.41
19.6.	30-40'	30-35 km	escort	Cornuda	Oblt. Pindter	138.43
					Stfw. Benko	138.44
					Kpl. Wissinger	38.42
	30-35'	30-40 km	fighter	Montello	Oblt. Pindter	138.43
					Kpl. Wissinger	38.42
					+ 2 pilots	
20.6	S: 6.45 L: 7.25	35-40 km	escort	Montello-Cornuda	Oblt. Pindter	138.43
					Stfw. Benko	138.44
					Kpl. Soviansky	38.30
					Kpl. Wesely	38.41

× this mission is not in Oblt. Pindter's log-book

Known Missions Flown by Pilots of Flik 74/J in June 1918

Date	Duration	Distance	Task	Route or territory of deployment	Pilot	Aircraft
21.6	50'-1h 25' S: 6.03 L: 7.15-7.30	35-45 km	defence	Cornuda	Oblt. Pindter	138.43
					Stfw. Benko	138.44
					Kpl. Wissinger	38.42
					Kpl. Wesely	38.30
	S: 16.00 L: 16.30				Kpl. Soviansky	38.30
					Kpl. Wissinger	38.42
	S: 18.27 L: 19.40	25-45 km	fighter	Montello (air fight, 1 unconfirmed victory)	Stfw. Benko	138.44
22.6	20'-57' S: 6.30-6.45 L: 6.50-7.42	15-30 km	defence	Conegliano	Oblt. Pindter	138.119
					Kpl. Soviansky	38.30
					Kpl. Staats	138.43
					Kpl. Wissinger	38.42
	10'-1h 8' S: 19.35 L: 19.45-20.43	35-45km	defence	Montello	Oblt. Pindter	138.119
					Stfw. Benko	138.44
					Kpl. Soviansky	38.30
					Kpl. Staats	138.43
					Kpl. Wissinger	38.42
23.6	50'-1h 25' S: 8.35 L: 9.55	35-45km	defence	Montello	Oblt. Pindter	138.119
					Kpl. Kunz	38.30
					Kpl. Soviansky	138.44
					Kpl. Staats (POW)	138.43
					Kpl. Wissinger	38.42
24.6	S: 16.50 L: 17.55	35 km	fighter	Montello	Oblt. Pindter	138.119
					Kpl. Wesely	138.44
	35'	35 km	fighter	Montello	Oblt. Pindter	138.119
25.6	30'	20 km	defence	Conegliano (air fight)	Kpl. Wissinger	38.42
28.6	S: 6.15-6.20 L: 7.22-8.03	40 km	escort	Cornuda	Oblt. Cserich	138.119
					Kpl. Kunz	38.27
					Kpl. Soviansky	138.44
					Kpl. Wesely	38.43
					Kpl. Wissinger	38.42

day, pilots of Flik 74/J (maybe along with ones of Flik 72/J) supposedly confronted with two Camels of No. 66 Squadron: *Capt.* W.G. Barker (B.6313) and *Lt.* G.F. Apps (B.7358). Barker described the duel as follows: *"Patrol observed eight E.A. Scouts at 7-5 a.m. at 16,000 ft over MORIAGO. Capt. Barker, with Lt. Apps, tried to surprise these E.A. but failed to do so, and E.A. dived towards GODEGA. Capt. Barker dived after E.A. and got on tail of a Berg Scout and after firing about 250 rounds E.A. fell out of control and broke into pieces at about 4,000 ft between CONEGLIANO and GODEGA. This was observed by Lt. Apps. In the meantime Lt. Apps engaged a D.III and after following E.A. down to about 5,000 ft, firing about 100 rounds, this E.A. turned over and fell to pieces just S. of GODEGA. This is confirmed by Capt. Barker. Capt Barker then attacked another D.III which was trying to attack Lt.*

Above: Aviatik D.I 138.43 in Italian hands. (Colin Overs via Boris Ciglić)

Apps. This E.A. passed in front of Capt. Barker at about 30 yards range, when Capt. Barker fired about 40 rounds and as E.A. dived pilot got on tail of E.A. and after firing about 100 rounds this E.A. stalled and went down out of control and crashed about one mile S. of GODEGA Aerodrome. This was seen by Lt. Apps. Capt. Barker then engaged a D.V. (with black & white checkered fuselage). This E.A. fought well and Capt. Barker was unable to get a good shot at E.A. After about 2 minutes manoeuvring Capt. Barker got in a burst at E.A. as E.A. was passing across of Capt. Barker's machine. E.A. immediately stalled and spun for about 4,000 ft and was seen to land on GODEGA Aerodrome. Patrol were then at 5,000 ft over GODEGA. All E.A. fought poorly except the last mentioned E.A."

That was typical combat with British overclaiming. The most interesting question is how British pilots could see two Austro-Hungarian fighters falling into pieces in the air and another one crashing when no casualties were reported on the opposite side. It is only known, that *Oblt.* Pindter fought with a "French type" at 3.000 meters altitude, and *Kpl.* Hauger's 38.27 got several hits in the encounter without any harm of the pilot.

As for the two-seater of Flik 59/D, Hansa-Brandenburg C.I 369.109 and its crew met some Sopwiths in the area of Sacile, probably on the way back, after leaving the escort fighters. The enemies were *Lt.* J. Mackereth (B.6344), *Capt.* J.E. Hallonquist (B.6363) and *Lt.* A.R. Strang (D.8209) of No. 28 Squadron: *"While on Long Offensive Patrol at 14,000 ft at 7-25 a.m. over PORDENONE, Lt Mackereth saw an enemy 2 seater flying at 11,000 ft. Lt Mackereth attacked the E.A. firing about 150 rounds into him and following him down about 3,000 ft. Lt Strang then fired a burst into him. Capt. Hallonquist re-attacked the E.A. and followed him down to 1,000 ft firing all the time and saw the machine crash near PORDENONE. Confirmed by Lt Mackereth."*

The pilot *Fwbl.* Cerwenka successfully force-landed with the strongly crippled plane at Ranzano (about 10 km W-NW of Pordenone), but his observer *Oblt. i.d.R.* Hans Reisenbauer got two headshots in the combat and was deceased shortly after landing.

Two days later, another accident happened when the wing ribs of *Zgsf.* Josef Marszalek's Aviatik D.I 38.63 broke. Marszalek crash-landed at Pianzano, on the airfield of Flik 47/F, where the plane was totally damaged, and its pilot was slightly injured.

On 16th July, three Camels of No. 66 Squadron, *Lt.* G.F.M. Apps (B.7358), *Lt.* A.E. Baker (D.9414) and

Above: The captured 138.43 was eventually displayed at Padua airfield on 20th September 1918 as a part of a great ceremony. (Paolo Varriale)

Lt. H.D. McDiarmid (D.8101), flew a close patrol at 13.000 feet when they observed an "LVG" escorted by four enemy scouts at 10.15 at 18.000 feet. The Austro-Hungarian fighters realised the foes, and according to the British report, they flew away. Patrol engaged the "LVG" and was shot down by *Lt.* Apps and *Lt.* Baker. The enemy reconnaissance aircraft was actually a Hansa-Brandenburg C.I of Flik 52/D and escorting fighters of Flik 74/J. The plane of pilot *Kpl.* Josef Krug and observer *Lt. i.d.R.* Karl Zinner was hit severely, but the crew remained unharmed and the pilot managed to land with the badly damaged aircraft. It is also known that *Oblt.* Salm von Hoogstraeten had a dogfight with three Sopwiths at the height of 400-900 meters, so the statement of running away of Flik 74/J pilots seems untrue.

In the morning of 23rd July, pilots of the squadron took off again for an escort mission of Flik 52/D. Over Conegliano-San Vendemiano Sopwith fighters of No. 28 Squadron engaged the Austro-Hungarian planes, and they managed to surprise and overcome them despite being inferior in number. *Capt.* T.F. Williams (D8242), *Lt.* A.G. Jarvis (D8170), and *Lt.* C.G. Constanduros (E1503) took off at 7.10 for a roving patrol and almost an hour late they discovered the enemy formation: *"While on Roving Patrol at 8-0 a.m. Capt. Williams saw 5 D.5's over PORDENONE Aerodrome, at about 14,500 ft. The E.A. proceeded towards the lines followed by the Camels, who remained unobserved until Capt. Williams attacked the leader at close range. A good burst resulted in the E.A. turning on its back and pieces of 3 ply flying off the fuselage. The Pilot appeared to right the machine and dived almost vertically with his engine off Lt. Jarvis then fired a good burst at 50 yards range. The E.A. turned on its back went down vertically and crashed about 2 km North East of S. POLO DI PIAVE. (38P.60.85)"*

According to Austro-Hungarian sources, seven own and five enemy planes took part in the dogfight, and one of the black-crossed fighters crashed burning at about 8.40-8-50 a.m. at Bocca near Cittadella, south of Conegliano and east of Santa Lucia. The remains of *Oblt.* Alexander Salm von Hoogstraeten's body were found in the wrecks of his Aviatik D.I 138.61. The other casualty of Flik 74/J was the commander, *Oblt.* Franz Cserich, whose fighter suffered serious damage, the upper wing of Aviatik D.I 138.116 broke, and he force-landed at Sacile. Cserich was injured and got a nerve shock, and his aircraft was ruined during the landing. *Kpl.* Soviansky force-landed with engine failure probably during or after this same mission.

Above & Below: Sopwith Camels B.6313 and B.6344, which played major roles in the British victories on 13th July 1918 (Andy Kemp)

Above: Officers of Flik 4 on the Wippach airfield in July 1917 (from left to right): Oblt. Mayer doctor, Lt. i.d.Res. Franz Janovsky, Lt. i.d.Res. Wilhelm Schandlbauer, Oblt. Johann Calogovič, Oblt. Rudolf Ortner, Lt. Arpad Pindter von Pindtershofen, Hptm. Alfons Veljačič CO, Lt. i.d.Res. Richard Schantin, unknown, Lt. Friedrich Wowy, unknown, Lt. i.d.Res. Emil Barkász. Oblt. Pindter served at Flik 74/J from 25th May 1918.

Below: Aviatik D.I 38.63 of Flik 74/J. Zgsf. Josef Marszalek crash-landed with this plane on 15th July 1918. (Rinaldo D'Ami via Paolo Varriale)

Left: Members of British No. 66 Squadron (from left to right): Lt. E.F. Mattock, Lt. T.F. Everett, Lt. C.G. Constanduros, Lt. R.B. Beevor (No. 28 Squadron), Lt. R.H. Foss Lt. Constanduros took part in the dogfight with pilots of Flik 74/J on 23rd July 1918. (Andy Kemp)

The squadron report at the end of July shows no satisfaction with the Aviatik D.I fighters, either. Series 138 and 115 were utilized as operational aircraft at that time and series 38 and 238 planes were as trainers. Although the flight characteristics of the first two types sufficiently met the requirements, some parts were not strong at all, particularly the wings proved very weak. The ribs of the upper wing on the exhaust side suffered highly from the exhaust fumes, and this was the reason for the bending and breaking of the rib tips on this side. The report mentioned three planes that were damaged due to wing failure: 138.54, 138.61, 138.62, (interestingly, the last two were originally assigned and flown by Flik 72/J), and four further fighters were as totally damaged: 38.30, 38.63, 38.65, 138.51. Another document mentions three fighters as written-off in July: 38.28 (after landing damage), 138.23 and 138.45, but these losses may have already occurred earlier.

Above: Oblt. Alexander Salm von Hoogstraeten, observer of Flik 3, who was killed in a dofight on 23rd July 1918 as a pilot of Flik 74/J. (Petr Aharon Tesař)

The month of August was uneventful as the unit remained in reserve because of the wing reinforcements on the "Bergs" being out of order. Six pilots were on leave after the exhausting two-monthly period, which caused high casualties. Moreover, many new members arrived at the squadron, especially from the already non-operational Flik 68/J and 72/J: four came in the last days of July and three more at the beginning of August.

By the order of the Austro-Hungarian Army High Command dated 20th August 1918 Flik 74/J became part of the newly established Jagdgeschwader of 6th Army under the command of *Hptm.* Alfred Dahinten, together with Flik 68/J and Flik 72/J which were already out of operational, without aircraft and personnel.

On the 22nd the unit still made practice flights, as the

Known Missions Flown by Pilots of Flik 74/J in July 1918

Date	Duration TO/Lndg	Distance	Task	Route or territory of deployment	Pilot	Aircraft
2.7	S: 7.30 L: 8.10	25-35 km	escort (Flik 59)	territory of II. Korps	Oblt. Cserich	138.119
					Oblt. Pindter	138.44
					Kpl. Kunz	38.27
					Kpl. Soviansky	38.30
					Kpl. Wesely	38.43
					Kpl. Wissinger	38.42
8.7	45' S: 9.30	40 km	defence	Conegliano–San Fior	Oblt. Cserich	138.119
					Kpl. Kunz	38.27
					Kpl. Soviansky	138.44
					Kpl. Wesely	38.43
					Kpl. Wissinger	38.63
13.7	1h 3-13' S: 6.37 L: 7.45	50 km	offensive patrol	Montello (air fight)	Oblt. Pindter	138.119
					Oblt. Salm-Hoogstraeten	138.52
					Kpl. Hauger	38.27
					Kpl. Marszalek	38.63
15.7	30-45' S: 7.30 L: 8.17	20-30 km	defence	Conegliano–Piave	Oblt. Pindter	138.119
					Kpl. Hauger	38.27
					Kpl. Marszalek (INJ)	38.63
					Kpl. Soviansky	138.44
					Kpl. Wissinger	38.66
16.7	1h 34' L: 11.05	65 km	escort (Flik 52)	Piave–Montello (air fight)	Oblt. Pindter	138.119
					Oblt. Salm	138.52
					Kpl. Wesely	38.27
	S: 9.30 L: 10.02	20 km			Kpl. Wissinger	38.66
17.7	1h 1-3'	50 km	offensive patrol	Montello–Mt. Tomba	Oblt. Salm	138.52
					Kpl. Wissinger	38.27
20.7	1h 21'	60 km	offensive patrol	Montello–Mt. Tomba	Kpl. Wissinger	138.55
23.7			escort (Flik 52)	Mt. Belluno–Treviso	Oblt. Salm (KIA)	138.52
					Oblt. Cserich (INJ)	138.116
					Kpl. Soviansky	
28.7	1h 15' S: 19.00	50 km	defence	Vittorio–Conegliano	Oblt. Pindter	138.119
					+ 6 pilots	
29.7	25'	20 km	defence	Pordenone–Sacile (air fight)	Oblt. Pindter	138.119

Above & Below: Aviatik D.I 138.55 *MIZZI* of Flik 74/J with different national markings: Eisernkreuz and Balkenkreuz. Kpl. Josef Kunz crashed the fighter plane shortly after take-off on 28th August 1918. (Martin O'Connor via Aaron Weaver)

Above: The personal marking of *MIZZI* on all surfaces of the fuselage in the white band on Aviatik D.I 138.55. (Colin Overs via Boris Ciglić)

Above: Aviatik D.I 138.116 was assigned to Flik 74/J and flown by squadron commander Oblt. Franz Cserich in the air combat of 23rd July 1918 when he force-landed with the aircraft. (Colin Overs via Boris Ciglić)

Above: The broken rib section of Aviatik D.I 138.119 after force-landing by Kpl. Adolf Wissinger on 19th September 1918.

formation of the squadron had not yet flown together. All missions of the month were carried out in the last week, but no air combat evolved during these fourteen sorties. On 28th August, however, a tragic accident happened when *Kpl.* Kunz crashed on the airfield shortly after take-off for a mission due to breaking the wing of his plane. The pilot and his Aviatik D.I 138.55 "MIZZI" completely burnt away because there was no available fire extinguisher.

Information on the daily life of Flik 74/J are more sporadic from September and October 1918, reports mention mostly accidents and other joyless events. At 6.20 p.m. on 4th September, an enemy aircraft dropped bombs on the neighbourhood of the airfield but caused no damage. On 19th September, *Kpl.* Wissinger made a practice flight in Aviatik D.I 138.119 when ribs of the left upper wing were broken. The pilot was able to force-land uninjured on the airfield through only desperate efforts.

During a night storm of 23/24 September, the unit lost two run-down hangar tents with two aircraft stored inside. Compared to other squadrons, Flik 74/J had a fortune because many units suffered disastrous damages that night.

The aircraft departed San Fior only two times in the last week of September. On the 25th, five pilots accomplished an uneventful fighter mission between 4.50 and 5.30 p.m. in the area of the Montello, while on the 29th, all three planes of the squadron, which were in operational condition, got involved in air combat with two enemy SIA aircraft south of Sacile. They took part in a defence sortie with all the available aircraft of Flik 42/J and 56/J, altogether seventeen planes. Due to the late alarm, however, the enemy could be no longer achieved. Only the three Berg fighters of Flik 74/J were able to reach the hostile planes far in the hinterland, but the fight remained scoreless because machine guns of all planes jammed owing to the bad ammunition.

There were significant movements among officer pilots in September. *Oblt.* Roman Schmidt became the new commander coming from Flik 30/J on the 4th, while his predecessor *Oblt.* Cserich was on leave from

Known Missions Flown by Pilots of Flik 74/J in August 1918

Date	Flt Time	Distance	Task	Route or territory of deployment	Pilot	Aircraft
25.8	55'	50 km	free hunting	Piave–Conegliano	Oblt. Pindter	138.119
					+ 3 pilots	
26.8	1h	65 km	free hunting	Sacile–Vittorio–Mt. Tomba–Piave	Oblt. Pindter	138.119
					+ 3 pilots	
28.8			fighter	Treviso–Volpago–Spresiano	Kpl. Kunz (KIA)	138.55
					+ 2 pilots	
31.8	1h 10'	70 km	free hunting	Sacile–Vittorio–Mt. Tomba–Piave	Oblt. Pindter	138.119
					+ 2 pilots	

1st August and should have been back on 4th September, but he was declared unfit for flying service as a result of his nerve shock suffered on 23rd July. *Lt. id.Res.* Heinrich Tarangul v. Valea-Utsai arrived on the 19th, and *Oblt.* Georg Vucetic on the 22nd, both directly from the fighter training at the field flying school. In the meantime, the most experienced officer pilot, *Oblt.* Pindter also went on leave between 17th September and 20th October, it is a question whether he returned to the unit until the end of the war, at all. *Oblt. i.d.Res.* Friedrich Bistritschan was transferred to Flik 74/J on 27th September from another squadron of San Fior and three days later left the unit his former companion at Flik 72/J, *Oblt.* Ferdinand Strommer.

Three days after the last deployment, the aircraft of the squadron took off from the airfield of San Fior on 3rd October for fighter and escort missions. Nine fighters of Flik 56/J and Flik 74/J flew in the area of the Montello, but the enemy plane could no longer be reached. On the same, day 11 single-seater of Flik 42/J and Flik 74/J were ordered to escort two reconnaissance planes of Flik 57/Rb, five of them, however, had to return – three due to engine failure. *Fwbl.* Juhász was one of the returning pilots, he crashed at Mareno, his Aviatik D.I 238.68 was totally damaged, the pilot was slightly injured.

On 6th October, the airfield of San Fior was bombed by the Italian airship M.14, coming from Ferrara to attack the base of Godega but the target was changed to San Fior because of the clouds. Bombs dropped from 4.000 meters, causing no harm. The unit carried out no missions until the 24th except on 5th October.

The last days of the month proved hectic. Opening events was an accident on the 24th, the first day of the entente offensive called Vittorio Veneto. *Fwbl.* Franz Birner (originally assigned to Flik 72/J) returned from a sortie and crashed during landing into a vineyard near the arfield of San Fior with Aviatik D.I 238.69. Some sources say that pilot remained uninjured, others stated he was fatally injured, but what is sure is that his aircraft was totally damaged.

On the next day, the strong rainfall inhibited any flying, but on the 26th the air came to life again despite the hazy and cloudy weather. Squadrons of the Austro-Hungarian 6th Army flew three reconnaissance and eight fighter missions. Pilots of Flik 74/J also took off and got involved in combat, where *Kpl.* Milan Soviansky was wounded in his Aviatik D.I 338.26. He managed to return to his own airfield and was carried into the hospital. Since, *Kpl.* Soviansky was the only known casualty of k.u.k. Luftfahrtruppe and *serg.* Marziale Cerutti of 79a Squadriglia the only one who claimed a fighter shot down that day, they probably faced each other. Cerutti escorted Caproni triplane bombers of 181a and 182a Squadriglia raided on Conegliano, and at 3.25 p.m., he engaged enemy fighters alone at the height of 2.800 meters. He described the fight as follows: *"Returning from an escort to Caproni triplanes, I saw between Mandre and Malanotte three enemy airplanes lower than me. I reached them and aimed well at the last one, firing about sixty rounds. Then I saw a large burst of smoke and flame, and the enemy airplane falling. I attacked the other two, but as my gun jammed, I had to leave the combat. Against the latter one I fired about 150 rounds."*

More than likely that several Aviatik D.I series 338 planes were assigned to the squadron, information about the operational history of these great fighters are very limited.

The day of 27th October brought intense air activity on both sides owing to especially the Italian raids carried out by Caproni squadrons. As the first action of the day, six three-engined planes of XI Gruppo (one Ca.3 of 4a Squadriglia and five Ca.5 of 6a Squadriglia) flew above

Above: Two Aviatik D.I fighters, a Phönix C.I and a Hansa-Brandenburg C.I reconnaissance aircraft, waiting for the inspection of the commanding officer of Austro-Hungarian 6th Army, Generaloberst Alois von Schönburg-Hartenstein, on San Giacomo airfield. The second one-seater from the left wears the squadron markings of Flik 74/J. (Martin O'Connor via Aaron Weaver)

Facing Page, Above: Kpl. Milan Soviansky in front of a "Berg" of Flik 74/J with a personal marking of a pipe-smoking skull on San Fior airfield. (Martin O'Connor via Aaron Weaver)

Facing Page, Below: Aviatik D.I series 38 or 138 of Flik 74/J with a personal marking of "gothic T". (Petr Aharon Tesař)

Left: Oblt. Roman Schmidt, commander of Flik 74/J from 4th September 1918.

Ni 3664

Above: Group photo of British No. 66 Squadron in Italy, 1918 (from left to right): back row: unknown, Lt. W. Robinson, Lt. L.R. Willis, Lt. C.M. Maud, Lt. F.S. Symondson, Lt. H.N.E. Row, Lt. R.G. Reid, unknown, Lt. H. Homan, Lt. Rudge, unknown, Lt. G.F.M. Apps, Lt. G.A. Birks, Lt. W.M. MacDonald; Middle row: Lt. J.S. Lennox, Capt. J.M. Warnock, Maj. J.T. Whittaker, Capt. H.B. Bell, Capt. P. Carpenter; seated on the ground: Lt. N.S. Taylor, Lt. S. Stanger, Lt. A. Jerrard, Lt. H.R. Eycott-Martin. Many of them fought against the pilots of Fliks 68/J, 72/J, and 74/J. (Andy Kemp)

Facing Page: Cap. Filippo Serafini and Ten. Giovanni Ancilotto were participants in the air battle over the airfield at San Fior on 28th October 1918. (Roberto Gentilli)

Right: Crash of another "Berg" fighter of Flik 74/J with a personal marking of "gothic T". The special camouflage pattern suggests a D.I produced by the W.K.F. factory. (Petr Aharon Tesař)

Sorties Flown by Pilots of Flik 74/J, May–September 1918

Name	Arrived	1918					Left
		May	Jun	Jul	Aug	Sep	
Oblt. Franz Cserich	30.3.1918	5	5	6	-	?	8.9.1918
Oblt. Arpad Pinter v. Pintershofen	25.5.1918	-	13	6	3	?	
Oblt. Alexander Salm v. Hoogstraeten	4.7.1918	-	-	4	-	-	23.7.1918 †
Lt. i.d.Res. Ernst Piech	12.4.1918	7	5	-	-	-	15.6.1918 †
Stfw. Karl Benko	21.5.1918	3	9	-	-	-	27.8.1918
Stfw. Johann Juhász	2.8.1918	-	-	-	2	?	
Fwbl. Johann Patleich	2.8.1918	-	-	-	3	?	
Zgsf. Josef Marszalek	4.7.1918	-	-	4	-	?	
Kpl. Adolf Wissinger	17.5.1918	-	9	6	-	?	27.10.1918 MIA
Kpl. Wenzel Wesely	17.5.1918	-	9	4	-	-	21.7.1918
Kpl. Milan Soviansky	17.5.1918	-	10	8	-	?	
Kpl. Josef Kunz	14.6.1918	-	3	8	1	-	28.8.1918 †
Kpl. August Staatz	21.6.1918	-	2	-	-	-	23.6.1918 POW
Kpl. Georg Hauger	11.7.1918	-	-	4	1	?	
Kpl. Karl Thomas	26.7.1918	-	-	1	-	-	22.8.1918
Kpl. Ladislaus Neftila	29.7.1918	-	-	-	4	?	
Altogether		15	65	51	14	67	

Vittorio Veneto to drop their bombs on ammunition depots without an escort; the fighters missed the planned rendezvous over Gazzo airfield. The bombers arrived over the target in loose formation, facilitating the work of the defenders, and during the encounters with enemy fighters on their way, two Capronis were lost. One of them was the only Ca.3, the place of its crash matches to the victory claimed by *Oblt.* Stefan Horváth of Flik 56/J. The other fell victim to *Oblt.* Roman Schmidt's machine guns, the commander of Flik 74/J, who claimed a Caproni shot down between Vittorio and Belluno at 12.30. The crew of Ca.5 11669 clashed with an Austro-Hungarian patrol of five fighters and were killed at Revine, northwest of Vittorio. The three-engined aircraft was flown by two American officers, *Lt.* DeWitt Coleman and *Lt.* James L. Bahl who were awarded Medaglia d'Oro al Valor Militare (Gold Medal for Military Valor). The justification for decoration of De Witt contains more details on the engagement:

"First Lieutenant of the American Aeronautical Corps, in the afternoon of 27 October 1918 during an action of bombing as crew commander of a Caproni aircraft, attacked by five enemy fighter, instead of escape the unequal struggle, landing, he preferred to accept it without hesitation, transfusing strength and energy to the flight companions with his magnificent example of resolve and courage. Two of the opponents were shot down by the infallible fire of the encircled plane, aboard of which was keep on fighting even amid the flames, until when, pressed and surmounted by the strong enemy patrol, it crashed and the whole crew paid with death for his audacity."

The observer of the bombers was *ten.* Vincenzo Cutello and the gunner was *serg.* Tarcisio Cantarutti.

Even though *Kpl.* Georg Hauger force-landed after a dogfight and *Kpl.* Adolf Wissinger did not return from a mission that day, it cannot be stated surely that they were lost in this duel.

Serious air combat occurred over San Fior on 28th October, but there is no available Austro-Hungarian report of it. Sopwith fighters of British No. 66 Squadron attacked withdrawn enemy troops and transport columns also on that day in many sorties. One of the flights consisted of *Lt.* A. Paget (E.7215), *Lt.* C.C. Butler (E.1497) and *Lt.* G.A. Goodman (E.1595) discovered

Flying Activities of Flik 74/J, May–September 1918

Period	At Last Day of Period		Flying Days Avail.	Flown			Fights	Vics	Losses in Combat
	Aircraft	Pilot		Sorties	Flt Time	Dist. (km)			
1-31. 5. 1918	5	9	13	15	18h 10'	1910	-	-	-
1-30. 6. 1918	8	8	18	65	57h	2115	3	-	-
1-31. 7. 1918	6	15	15	51	44h 10'	1915	4	-	1
1-31. 8. 1918	-	5	8	14	15h 45'	835	-	-	-
1-30. 9. 1918	6	11	22	67	64h	7990	4	-	-

Flown sorties according to the half-monthly reports: 15–31.5.1918: 14 (17h 25'), 1–14.6.1918: 20 (23h 14'), 15–30.6.1918: 56 (46h 12'), 1–15.7.1918: 18 (16h 9'), 16–31.7.1918: 25 (35h 42'), 16–31.8.1918: 8 (8h 46'), 1–15.9.1918: 27 (28h 36')

thirty rolling stock on the railway leading north from Visnadello. *Lt.* Paget and *Lt.* Butler dropped their bombs on this target, but *Lt.* Goodman did not go down with them. The patrol then observed what looked like a Camel and two enemy aircraft fighting over Godega airfield at about 1,000 ft at 8.50 a.m. The machine taken for a Camel was smoking and eventually went down in flames and crashed east of the airfield. *Lt.* Paget engaged the foes identified as Albatros D.Vs, and according to their report, he shot down both. Italian fighters also joined this duel, and despite their combat report was dated 27th October, its details perfectly fit the event. *Cap.* Filippo Serafini, commander of 77ª Squadriglia and *ten.* Giovanni Ancilotto arrived late to the scene, and they witnessed only the fall of *Lt.* Goodman and his plane:

"Time of departure: 8.25
Time of landing: 9.25
Shots fired: 500
Type of attacked enemy aircraft: Pfaltz fighter
Time, height and place of combat:
Way of attack: time 9 – 1,000 meters – S. Fior airfield, from behind and below Cap. Serafini, from behind and above Ten. Ancilotto followed in a vertical dive down to 200 meters from the ground.
The exact place of the shot-down aircraft's crash: near the airfield.
Other airplanes that took part in this combat: two Englishmen who engaged in combat with three more enemy fighters.
Additional notes: we took off for a free hunt flight and flew to S. Fior airfield. We saw an enemy attacking a British fighter. Immediately Ten. Ancilotto rushed to defend it, but he failed in saving it as it caught fire. Then he attacked the enemy and saw diving vertically while being hit and followed it down to 200 meters. He considers it shot down. Capitano Serafini meanwhile attacked a Pfaltz at a height of 800 meters and this was shot down and crashed, as witnessed also by Ten. Ancilotto. Meanwhile, two more British fighters and three enemy fighters arrived and engaged in combat. Two aircraft went down in flames, but we could not identify their nationality."

Austro-Hungarian participants and casualties are not known, but the victorious pilot was possibly *Fwbl.* Johann Juhász, that could be the only reasonable explanation why he was awarded Golden Bravery Medal after the war for his service at Flik 74/J.

With the retreat of Austro-Hungarian troops, the camp of San Fior di Sopra was also evacuated by 30th October. *Serg. magg.* Luigi Capparucci of 78ª Squadriglia was hit during a mission that day and forced to land on the airfield while his companion *serg.* Oreste Codeghini covered him, landing close to this place. Codeghini proved a great help as he freed Capparucci from an Austrian gendarme and took his companion home hanging on his Hanriot fighter. The same day but later, the ace *ten.* Ferruccio Ranza of 91ª Squadriglia also force-landed on the airfield of San Fior. He came from a strafing mission with a hit in his tank, and he experienced that it was abandoned by the Austrians, so he could easily refill his plane with Austrian gasoline and returned to his base.

Colors & Markings

It is not the goal of this chapter to examine this large subject in great depth, simply to give insight by the limited amount of information.

As for Flik 68/J, series 153 fighters of the Oeffag-built Albatros D.III were the first planes that were allocated to the squadron. Since almost all aircraft of this series left the factory without any camouflage, the first (or most of) single-seaters of the squadron were probably also delivered in the same way, bearing material colours, i.e. varnish linen, plywood, and riffled sheet. From March 1918, the aircraft factories were increasingly urged to apply camouflage on all planes, particularly on D.IIIs. In response to the demand, a fabric of a special swirled pattern was produced by the Oeffag factory, covering partly (wings, tailplane, rudder) the aircraft with it from April 1918. In summer 1918, Oeffag applied a camouflage of dabbed mottles on several Albatroses, mostly of series 253. The main colours of this pattern were shades of green (dark green, grey-green, grey-yellow). Operational planes were received field camouflages in many cases, already from autumn 1917. These were dabbed dark camouflage (of one or two shades of dark green) or similar to the factory pattern with dabbed mottles of green shades, applying on the upper and lateral surfaces.

According to the only known photo of the squadron's operational aircraft, one of Stfw. Andreas Dombrowski's crash-landed 153.195, this plane had a fuselage and upper surfaces painted with a dark shade of colour, and under surfaces with light or material colour.

The reconstruction of the exact camouflage patterns and personal markings of the squadrons' Albatros fighters is not possible due to the lack of photographic evidence, similarly to the Phönix fighters of Flik 68/J. Colours of Phönix D.II/D.IIa fighters are suggested by the statements of former Austro-Hungarian pilots, Julius Arigi and Mario Tschurtschenthaler, who reminded the type as being painted basically green. Tschurtschenthaler was a member of Flik 21, stationed at Pergine and he described the camouflage of Phönix D.IIa 422.08: "It carried scruffy green camouflage. All its surface was covered with mottles of green." This painting was usually applied not only on all the fuselage, the fin and rudder, and the upper surfaces of the wings and the elevators, but also on wing struts, undercarriage legs and also often wheel discs. The under surfaces of the wing and the elevators its cover remained in the natural colour of the material.

Based on the only revealed personal marking on 222.04 (or more likely 222.08), the letter "X", the pilots of the squadron possibly used also letters on other Phönix D.II/D.IIa fighters as markings.

The unit marking of Flik 68/J is known from a document of the 6th Army dated 14th April 1918: aircraft were provided green-white wheel discs. This was probably applied to the fighters already in spring 1918, however, there is no available evidence of it.

Flik 72/J and Flik 74/J were equipped exclusively with Aviatik D.I fighters, which were produced by three different factories (out of the six which manufactured this type). That also meant divergent patterns of camouflage. The first camouflage pattern, named "Autumn Leaf Mottle" by Martin O'Connor, was applied in autumn 1917 and consisted of mottles and splotches in two or three colours painted with rags or sponges. A variant of this camouflage pattern was used on a plane of Flik 74/J: the pattern of little mottles dubbed on the base colour, with sporadic smudges of "streaked pattern".

Beginning with the spring of 1918, "Berg" fighters received a camouflage pattern of handpainted hexagons (lozenge), usually arranged in light and dark colour bands, running diagonal or vertical from nose to tail. An exception to this arrangement can be seen among the aircraft of Flik 74/J, which seems to have worn only a light colour pattern.

Aviatik basically employed almost equilateral hexagons in two lozenge patterns ("edge down" and "point down"), while the other company, Lohner, applied two schemes on its Aviatik fighters: one with equilateral hexagons and the other with smaller rectangular polygons, flattened hexagons (also in "edge down" and "point down" patterns). The "edge down" hexagons were usually used on the fuselage, and fin and rudder, and the "point down" ones on the wing and elevator. What was common to all Aviatik D.Is, that camouflage was utilized on upper and side surfaces only, while lower surfaces of wings, fuselage, and tailplane remained in plain finish.

Aviatik D.I fighters carried unit markings and personal insignia by the common standards of the Austro-Hungarian aviation troops (k.u.k Luftfahrtruppen), as well. Flik 72/J had its aircraft provided with a black stripe on the fuselage and black wheel discs. White numbers inside the stripes served for the identification of individual aircraft. Flik 74/J marked its aircraft with a white stripe behind the cockpit and white wheel discs. Black letters, various emblems and inscriptions, exceptionally numbers facilitated prompt identification.

The detailed Italian and British reports on captured 138.43 described its marking as follows: Black "T" on a white ground on the side and upper facings of the fuselage, midway between the tail and the pilot's seat." Accordingly, the assumption that the features were usually painted on all four surfaces of the fuselage is not necessarily correct, or not in every case, at least.

As usual for the Austro-Hungarian aircraft, the national insignia was the black Iron Cross (Eisernes Kreuz) outlined in white. Planes beared two large black crosses on the upper wings, two under the lower wings, and one on each side of the rudder. A memorandum, issued on 31st July 1918, suggested replacing Iron Crosses with straight-armed crosses (Balkenkreuz), but it did not become generally applied until the end of the war.

Right: A poor quality but unique photo of Aviatik D.I 338.95 with a special personal emblem and late-war national markings. This aircraft is believed to possibly served at Flik 74/J. (Reinhard Masajdek via Petr Aharon Tesař)

Below: Aviatik D.I 38.03 at Aspern, assigned later to Flik 74/J for a short time. (Martin O'Connor via Aaron Weaver)

Color Profiles

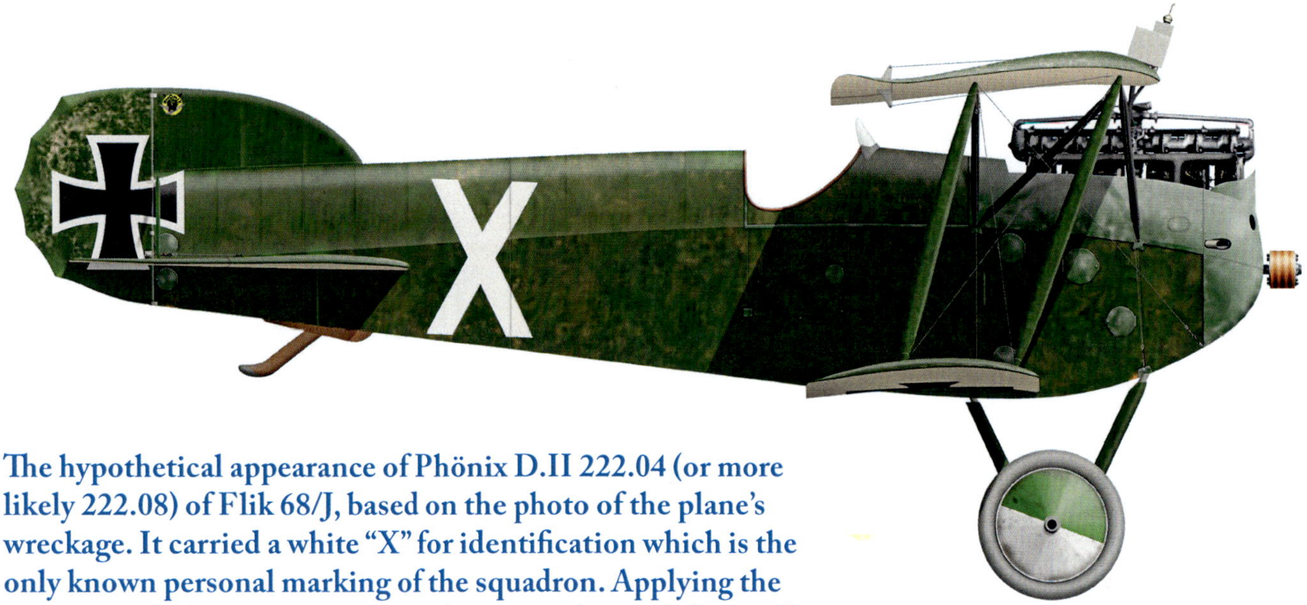

The hypothetical appearance of Phönix D.II 222.04 (or more likely 222.08) of Flik 68/J, based on the photo of the plane's wreckage. It carried a white "X" for identification which is the only known personal marking of the squadron. Applying the unit marking (green-white wheel discs) on Phönix fighters of Flik 68/J is questionable.

Aviatik D.I (Lo) 115.25 of Flik 72/J. Fwbl. Ernst Heinz crashed with the aircraft on 2nd June 1918. The series 115 fighter had a typical camouflage pattern of lozenge of flattened hexagons. Unlike the Aviatik-built aircraft, the "Bergs" produced by Lohner had camouflaged engine panels. The white number in a black band was a unit marking for identification, the exact number ("2") is uncertain in this case (maybe"6").

Aviatik D.I 138.62 shows the squadron markings of Flik 72/J, a black band on the fuselage with a white number and the black wheel discs. The aircraft crashed after a fight on 29th July 1918, killing its pilot Oblt. August Dehne.

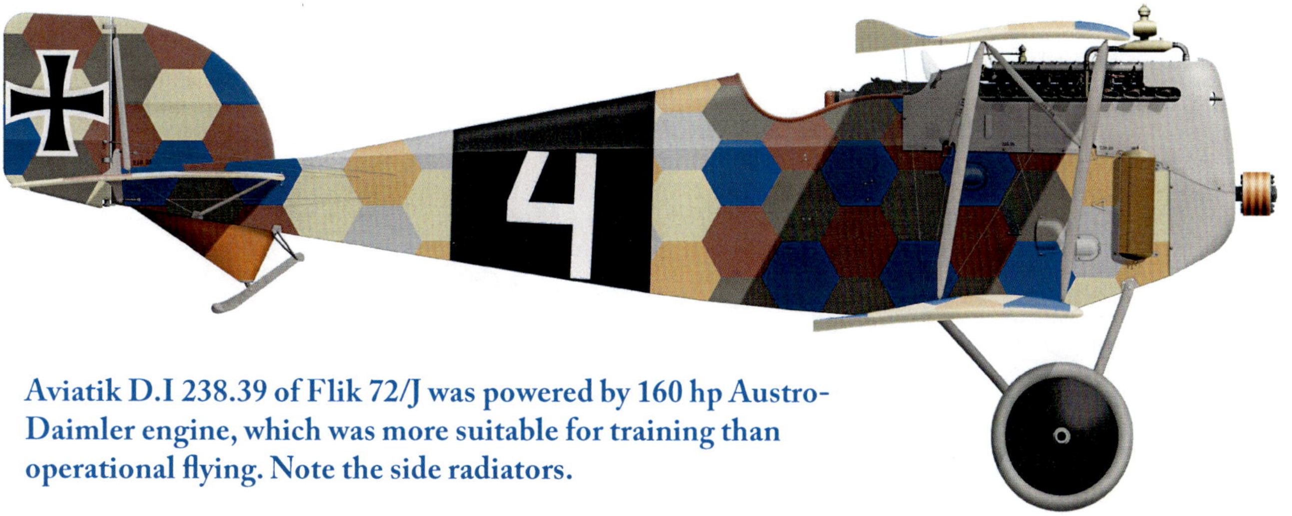

Aviatik D.I 238.39 of Flik 72/J was powered by 160 hp Austro-Daimler engine, which was more suitable for training than operational flying. Note the side radiators.

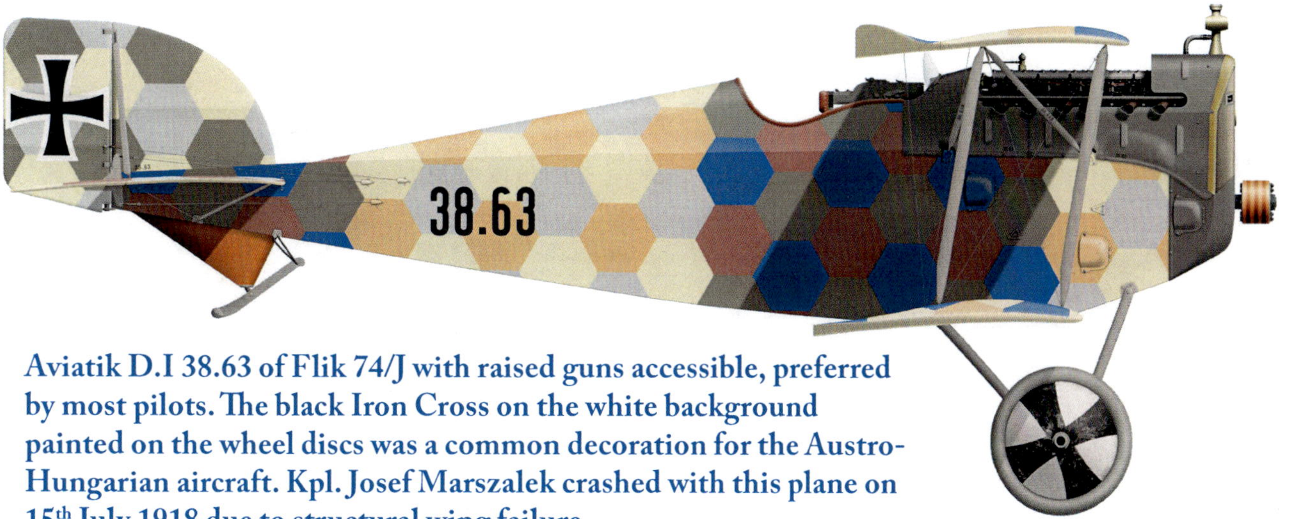

Aviatik D.I 38.63 of Flik 74/J with raised guns accessible, preferred by most pilots. The black Iron Cross on the white background painted on the wheel discs was a common decoration for the Austro-Hungarian aircraft. Kpl. Josef Marszalek crashed with this plane on 15th July 1918 due to structural wing failure.

Aviatik D.I 138.43 flown by Kpl. August Staatz of Flik 74/J was forced to land near Treviso on 23rd June 1918. The personal marking ("T") is believed to have belonged to Kpl. Karl Thomas, but he arrived at the unit on 26th July 1918.

Aviatik D.I 138.55 before and after having received new national markings in August 1918. On the latest version, the personal marking ("MIZZI") was applied on the upper and maybe also on the under surfaces of the wide white band, while the earlier version beared it only on the sides. The aircraft was damaged by Kpl. Josef Kunz, crashing shortly after take-off on 28th August 1918.

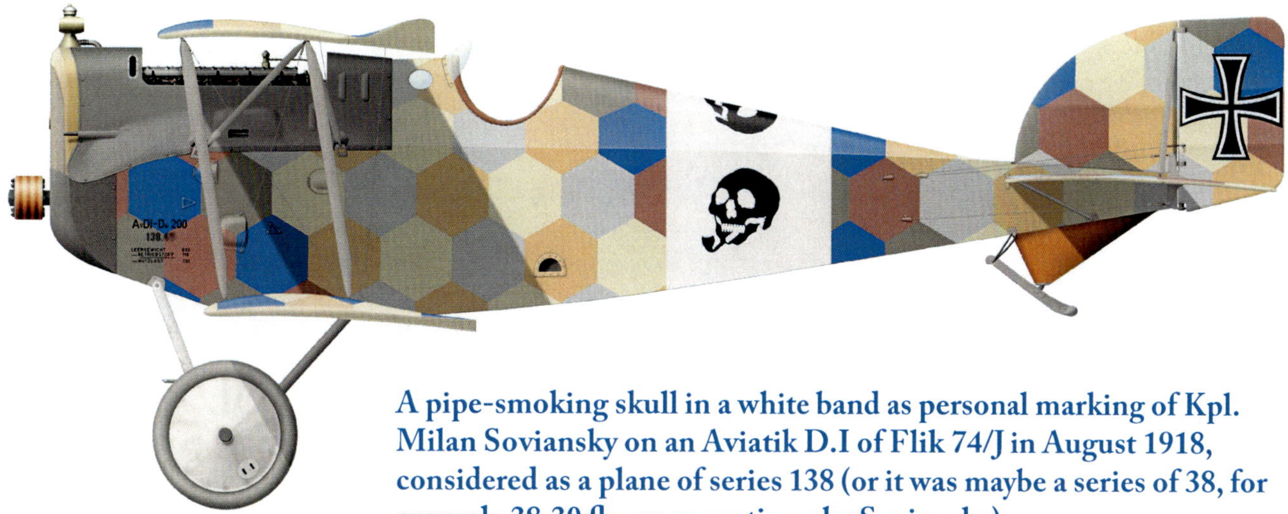

A pipe-smoking skull in a white band as personal marking of Kpl. Milan Soviansky on an Aviatik D.I of Flik 74/J in August 1918, considered as a plane of series 138 (or it was maybe a series of 38, for example 38.30 flown many times by Soviansky).

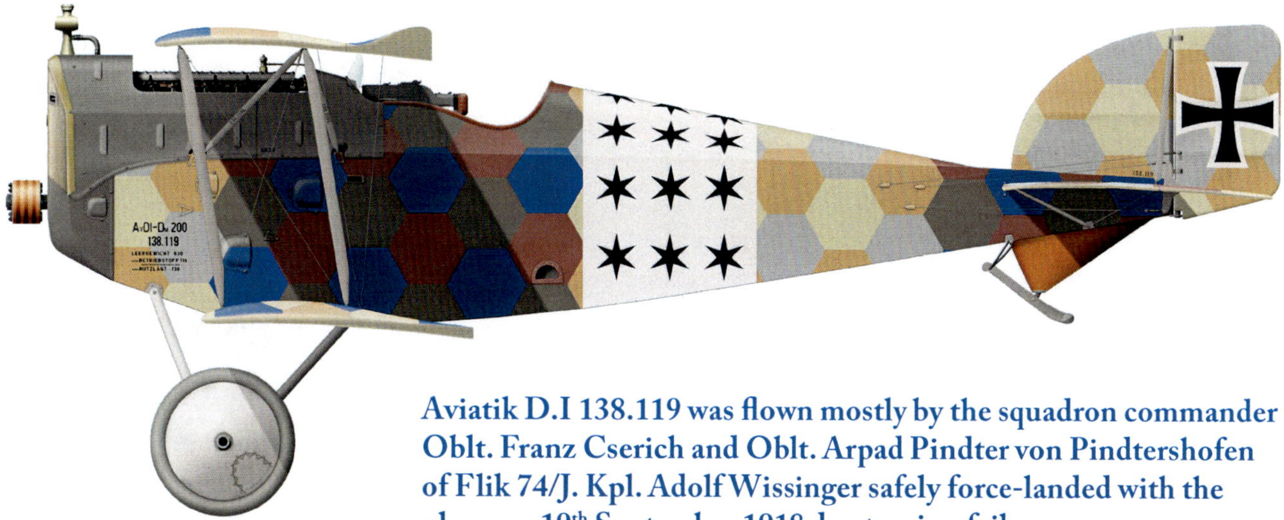

Aviatik D.I 138.119 was flown mostly by the squadron commander Oblt. Franz Cserich and Oblt. Arpad Pindter von Pindtershofen of Flik 74/J. Kpl. Adolf Wissinger safely force-landed with the plane on 19th September 1918 due to wing failure.

Aviatik D.I series 38 or 138 of Flik 74/J with a unique camouflage pattern and a personal marking (gothic "T").

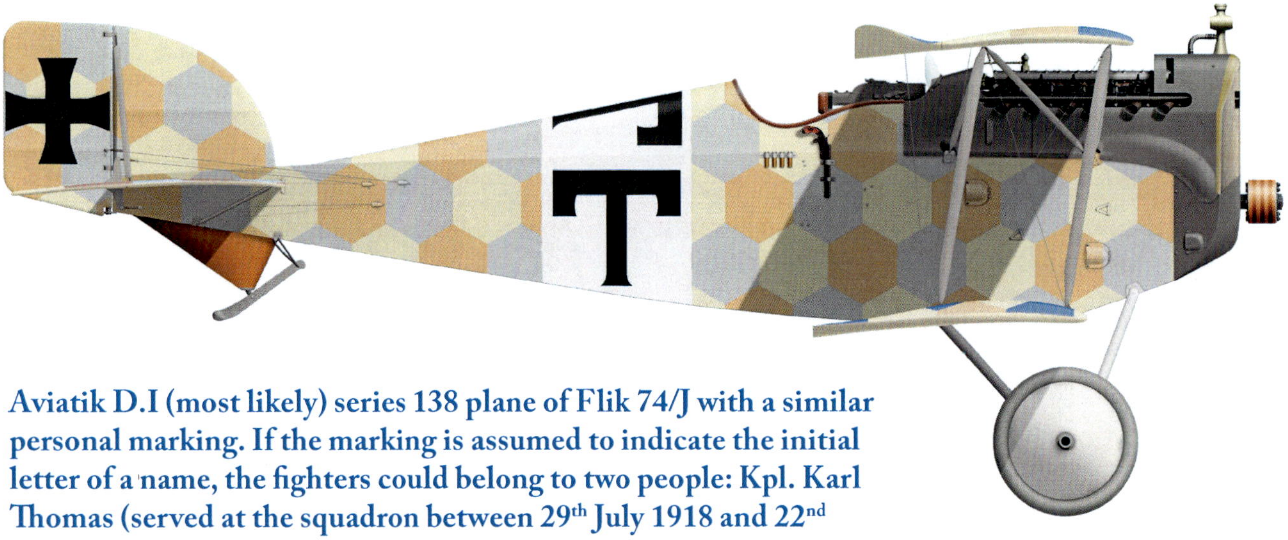

Aviatik D.I (most likely) series 138 plane of Flik 74/J with a similar personal marking. If the marking is assumed to indicate the initial letter of a name, the fighters could belong to two people: Kpl. Karl Thomas (served at the squadron between 29th July 1918 and 22nd August 1918) or Lt. i.d.Res. Heinrich Tarangul v. Valea-Utsai (arrived at the unit on 22nd September 1918).

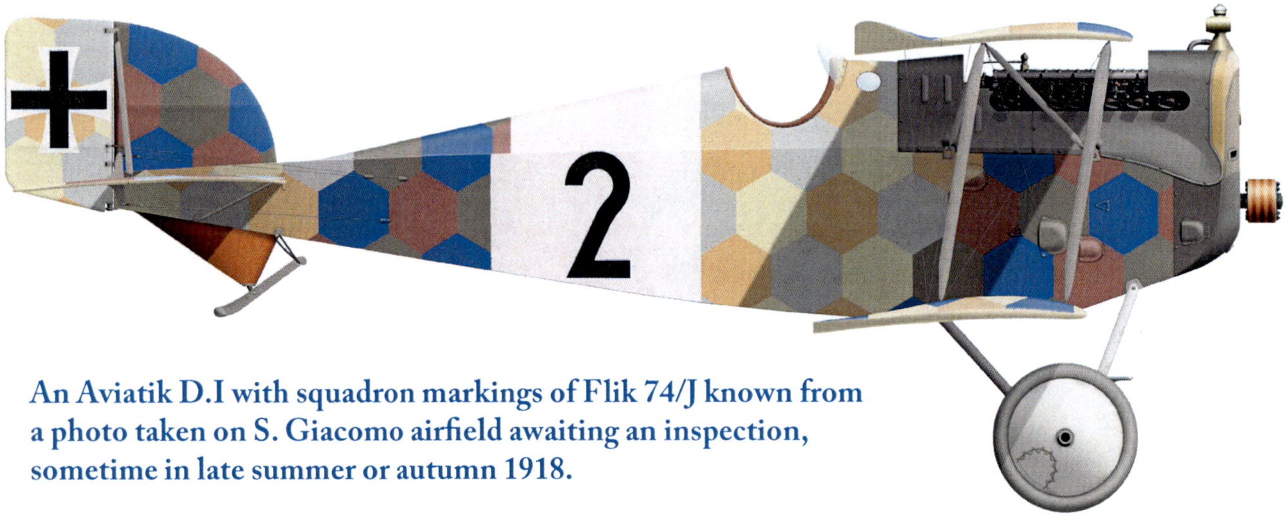

An Aviatik D.I with squadron markings of Flik 74/J known from a photo taken on S. Giacomo airfield awaiting an inspection, sometime in late summer or autumn 1918.

Appendices

Air Victories

Flik 68/J

Date	Pilot (Number of Victory)	Aircraft Used	Area	Enemy	Note
4.5.1918	Stfw. Andreas Dombrowski (6)	Albatros D.III (Oef) 153.195	Montebelluna	Sopwith single-seater	U/I
	Kpl. Franz Fritsch (1–2)	Albatros D.III (Oef) 153.210	enemy territory	2 enemy single-seaters	U/I

Flik 72/J

Date	Pilot (Number of Victory)	Aircraft Used	Area	Enemy	Note
21.6.1918	Oblt. i.d.Res. Dr. Josef Hoffmann von Ostenhof (-)	Aviatik D.I 138.61	Nervesa	Nieuport single-seater	U/I

Flik 74/J

Date	Pilot (Number of Victory)	Aircraft Used	Area	Enemy	Note
15.6.1918	Stfw. Karl Benko (4)	Aviatik D.I 138.44	at the Montello, enemy territory	SVA or SPAD single-seater	U/I
21.6.1918	Stfw. Karl Benko (-)	Aviatik D.I 138.44	Trevignano enemy territory	Sopwith single-seater	U/C U/I
25.6.1918	Kpl. Adolf Wissinger (-)	Aviatik D.I 38.42	enemy territory	SPAD single-seater	U/C U/I
27.10.1918	Oblt. Roman Schmidt (6)	Aviatik D.I	Vittorio–Belluno	Caproni Ca.5 Ca.11669, 6a Sq., Lt. Coleman F. De Witt Jr., Lt. James L. Bahl, ten. Vincenzo Cutello, serg. Tarcisio Cantarutti (all KIA)	

Aircraft Strength

Flik 68/J

30 May–28 June 1918

Date	30.5–1.6	14.6.	16.6.	17.6.	18.6.	19.6.	
Ready to Start	153.101 153.113 153.192* 153.194* 153.205 238.42	153.261 222.06 222.07 222.08 422.26	222.06 222.07 422.26	222.07 222.08 422.26	222.06 222.07 222.08 422.26	222.06 222.07 222.08	
In Repair	153.256* 153.257 153.258* 153.259* 153.260* 153.261	238.42	222.08	222.06 222.14 422.02 422.19 422.24	222.04 222.14 422.02 422.19 422.24	222.04 422.02 422.19 422.24 422.26	
Damaged			238.42	238.42			

Date	20-21.6	22.6	23–24.6	25.6	26.6	27–28.6	29.6
Ready to Start	222.08 422.19	222.04 422.19 422.24	222.04 422.19 422.24 153.205¤	222.04 422.02 422.19 422.24 153.205¤	222.04 422.02 422.19 422.24	422.02 422.03 422.19 422.24	422.02 422.03 422.19 422.24
In Repair	222.04 422.02 422.03 422.24 422.26	422.02 422.03 422.26	422.02 422.03 422.26	422.03	422.03	222.04	222.04 222.06
Damaged	153.205 222.06 222.07 222.14	153.205 222.06 222.07 222.08 222.14	222.06 222.07	222.06 222.07× 422.26	153.205 222.06 222.07× 422.26	222.06 222.07×	222.07×

* handed over to Flik 30/J on 10.6.1918
¤ training aircraft
× without engine

Flik 72/J
17–28 June 1918

Date	17-19.6	20.6	21.6	22.6	23.6	24.6	25.6	26-28.6	30.6
Ready to Start	115.31 115.10 138.61 138.62 184.01	115.31 138.61 138.62 184.01	115.10 138.61 138.62 184.01	184.01 238.54	184.01 238.54	238.54	238.54	138.61 138.62	138.61 138.62 115.10 115.35
Training Aircraft	238.23 238.54 46.66	238.23 238.54 46.66	238.23 238.54 46.66	238.23 46.66	238.23 46.66	238.23 46.66	238.23 46.66	238.23 238.54 46.66	46.66
In Repair	115.54× 238.24	115.10 115.54× 238.24	115.02× 115.18× 115.31 115.35× 115.54× 238.24	115.02× 115.10 115.18× 115.35× 115.54× 138.61	115.02× 115.10 115.18× 115.35× 115.54× 138.61 138.62 238.24	115.02× 115.10× 115.18× 115.35× 115.54× 138.61 138.62 184.01	115.02× 115.10× 115.18× 115.35 115.54× 138.61 138.62 238.24 184.01×	115.02× 115.10× 115.18× 115.35 115.54× 238.24 184.01×	115.54 138.46 138.49
Damaged			238.42	238.42					

× without engine

Flik 74/J
21–23 June 1918

Date	16.6	21.6	22.6	23.6
Ready to Start	138.43 138.44 138.119	138.43 138.44	138.43 138.44 138.119	138.43 138.44 138.119
Training Aircraft	38.30 38.41 238.34	38.27 38.30 38.42 238.34	38.30 38.42 238.34	38.30 38.42 238.34
In Repair	38.03 38.42 38.43 138.09 238.45	38.03 38.41 38.43 138.09 138.119 238.45	38.03 38.27 38.41 38.43 138.09 138.51 238.45	38.03 38.27 138.09 138.51 238.45
Arrived at the Squadron		138.29 138.51 138.52		
Reserved without Engine			138.29 138.52	

Known Aircraft in Use

(with engine number in brackets)

Flik 68/J

Albatros D.III (Oef)	Aviatik D.I	Phönix D.II/D.IIa
153.43 (19197)	38.04 (?)	222.04 (33717)
153.101 (19126)	38.31 (?)	222.06 (33701/33145)
153.108 (19257)	238.42 (17601)	222.07 (33686)
153.113 (19228)		222.08 (32628)
153.147 (19299)		222.14 (33769)
153.176 (19338)		422.02 (34463)
153.177 (19344)		422.03 (34474)
153.182 (19372)		422.04 (?)
153.183 (19366)		422.19 (34481)
153.184 (19086)		422.24 (34485)
153.189 (19162)		422.26 (34466)
153.192 (19321)		
153.193 (19378)		
153.194 (19374)		
153.195 (19341)		
153.196 (19373)		
153.205 (19391)		
153.210 (19380)		
153.256 (19503)		
153.257 (19530)		
153.258 (19501)		
153.259 (19529)		
153.260 (19548)		
153.261 (19299)		

Left: Crash of Aviatik D.I (Lo) 115.51 at an unidentified unit and on an unknown date. The sixth from the left behind the aircraft is Zgsf. Ignaz Pillwein who served also at Flik 68/J. (Martin O'Connor via Aaron Weaver)

Flik 72/J

Aviatik D.I				Lloyd C.V
115.02 (19237)	138.16 (1914?)	238.05 (?)	184.01 (19209)	46.66 (81100)
115.03 (19505)	138.36 (19253)	238.22 (?)		
115.10 (19510)	138.46 (19360)	238.23 (17516)		
115.12 (19512)	138.49 (19387)	238.24 (18360)		
115.13 (19609)	138.61 (19214)	238.39 (17337)		
115.15 (19589)	138.62 (19394/19414)	238.54 (17580)		
115.18 (19526)	138.98 (19349)			
115.25 (19521)				
115.31 (19533)				
115.35 (19538/19209)				
115.54 (19568)				

Flik 74/J

Aviatik D.I				
115.13 (19609)	38.03 (18170)	138.09 (19213)	238.07 (18073)	338.26 (23087)
	38.27 (18248(2))	138.23 (19101)	238.34 (17102)	
	38.28 (18345)	138.29 (19192)	238.45 (1709?)	
	38.30 (18189)	138.36 (19253)	238.68 (17525)	
	38.41 (18258)	138.43 (19351)	238.69 (17611)	
	38.42 (18089)	138.44 (19357)		
	38.43 (18175)	138.45 (19354)		
	38.63 (18095)	138.51 (19384)		
	38.65 (18221)	138.52 (19006)		
	38.66 (18032)	138.54 (19405)		
		138.55 (19408)		
		138.61 (19214)		
		138.63 (?)		
		138.116 (?)		
		138.119 (19367)		

ARCHIVAL SOURCES

ÖSTERRECHISCHES STAATSARCHIV/KRIEGSARCHIV (WIEN)
Luftfahrtarchiv
- Tagesbefehle der k. u. k. Luftfahrtruppen, 1916–1918
- Personalakten
- Frontlagemeldungen
- Tagesberichte
- Monatsberichte der Fliegerkompagnien (Flik 68/J, Flik 72/J, Flik 74/J)
- Monatsberichte der Fliegerkompagnien an die Flugzeuggruppe
- Technische Monatsberichte der Fliegerkompagnien an die Flugzeuggruppe
- Tagesbefehle des k.u.k. Fliegerarsenals
- Luftkämpfe
- Luftsiege
- Verluste
- Halbmonatsberichte der 6. Armee (Koluft Berichte)

Neue Feldakten
- Operativ Akten (II., XXIV. Korps)

Belohnungsanträge

NATIONAL ARCHIVES (LONDON)
- Combat in the air reports (14th Wing)
- Combat Reports (No. 28, 45 Squadron)
- Record Book (No. 28, 45, 66 Squadron)
- War Diary (7th Brigade, 14th Wing)

HONVÉDELMI MINISZTÉRIUM HADTÖRTÉNELMI INTÉZET ÉS MÚZEUM, HADTÖRTÉNELMI LEVÉLTÁR (BUDAPEST)
Tiszti anyakönyvek
Kitüntetési javaslatok
I. vh. hadosztályparancsnokságok iratai (M. kir. 11. honv. lovashadosztály)

KÖZLEKEDÉSI MÚZEUM ARCHÍVUMA (BUDAPEST)
Kézirattár
- Bakonyi Rudolf repülő visszaemlékezése

Okmánytár
- Pilóta-igazolvány a Nemzetközi Léghajós Szövetségtől Neftila László részére (1918)

ARCHIVIO UFFICIO STORICO STATO MAGGIORE AERONAUTICA (ROMA, VIA ROBERTO GENTILLI)
Fondo Prima Guerra Modiale
– Fondo Squadriglie, Cartella 76ª Squadriglia, Diario Storico 76ª Squadriglia Caccia

BIBLIOGRAPHY

Callegari, Renato: *I nidi delle aquile. Campi di aviazione della grande guerra nel Veneto.* Stamperia Cetid, Venezia, 2015.

Di Martino, Basilio: *L'Aviazione italiana e il bombardamento aereo nella Grande Guerra.* Ufficio Storico Aeronautica Militare, Roma, 2013.

Gentilli, Roberto – Iozzi, Antonio – Varriale, Paolo: *Italian Aces of World War One and Their Aircraft.* Schiffer Publishing, Atglen, 2003. 495 p.

Gentilli, Roberto – Varriale, Paolo: *I Reparti dell'aviazione italiana nella Grande Guerra.*

Ufficio Storico Aeronautica Militare, Roma, 1999.

Grosz, Peter M. – Haddow, George – Schiemer, Peter: *Austro–Hungarian Army Aircraft of World War One.* Flying Machines Press, Mountain View, California, 1993.

Grosz, Peter M.: Aviatik D.I. *Windsock Datafile 45.* Albatros Productions, Berkhamsted, 1992.

Grosz, Peter M.: Phönix D.I-D.II. *Windsock Datafile 31.* Albatros Productions, Berkhamsted, 1992.

Heubel, Johannes – Meindl, Karl: *Die Luftsiege der k. u. k. Luftfahrtruppen. Italienische Front. 1. Teil, 1918.* Haid, 2005.

Meindl, Karl: *Die Luftsiege der k. u. k. Luftfahrtruppen. Italienische Front. 2. Teil, 1918.* Haid, 2007.

Meindl, Karl – Schiemer, Peter: *Die Verluste der k.u.k. Luftfahrtruppen und ihre Luftsiege (unter Einschluß deutscher Flieger an der italienischen Front).* 1996. (unveröffentlicht)

Nachrichtenblatt der k.u.k. Luftfahrtruppen. Kommandant der k.u.k. Luftfahrtruppen, I-II. Jahrgang. Wien, 1917-1918.

O'Connor, Martin: *Air Aces of the Austro–Hungarian Empire, 1914–1918.* Champlin Fighter Museum, Mesa, 1986.

Scaroni, Silvio: *Battaglie nel cielo.* Longanesi & C., Milano, 1971.

Schiemer, Peter: *Die Albatros (OEFFAG). Jagdflugzeuge der k.u.k Luftfarhrtruppen.* H. Weishaupt Verlag, Graz, 1984.

Schroeder, Walter: *Die Geschichte der Fliegerkompagnie 74J.* ÖFH Nachrichten 2/2005. p. 4-14.

Schroeder, Walter – Tötschinger, Bernhard: *Luftsiege der Flik 74. 1914-1918.* ÖFH Nachrichten Sonderheft 21. Wien, 1993.

Soré, Piero: *La guerra aerea da Caporetto a Vittorio Veneto. 24 ottobre 1917 – 4 novembre 1918.* Aviani & Aviani editori, Udine, 2015.

Tesař, Petr Aharon: *Albatros D.II & D.III OEFFAG.* JaPo, Hradec Králové, 1998.

Varriale, Paolo: *I caduti dell'aviazione italiana nella Grande Guerra.* Ufficio Storico Aeronautica Militare, Roma, 2014.

Varriale, Paolo: *Aviatik (Berg) D.I at War! Windsock Datafile 171.* Albatros Productions, Berkhamsted, 2016.

Veinfurter, Robert: Das Fliegende Personal der k.u.k. Fliegerkompagnien im Ersten Weltkrieg. *ÖFH Sonderband 34.* Wien, 2018.

Zahálka, Jan – Tesař, Petr Aharon – Zdeněk, Skolil: *Aviatik D.I & D.II.* JaPo, Hradec Králové, 2002.

Zahálka, Jan – Tesař, Petr Aharon – Tyrlik, Sigmund: *Phönix D.I – D.III.* JaPo, Hradec Králové, 2005.

ABBREVIATIONS

A – Artillerie
cap. – caporale
Capt. – Captain
CO – commanding officer
D – Division (the original meaning, but later signed a general-purpose unit)
E.A. – enemy aircraft
F – Fernaufklärung
FA – Fliegerabteilung
Flik – Fliegerkompagnie
Fwbl. – Feldwebel
Hptm. – Hauptmann
i.d.Res. – in der Reserve
INJ – injured
J – Jagd
Jasta – Jagdstaffel
KIA – killed in action
KIAcc – killed in accident
Kpl. – Korporal
Lt. – Leutnant/Lieutenant
Lo – Lohner
magg. – maggiore
Maj. – Major
MMKM – Magyar Műszaki és Közlekedési Múzeum (The Hungarian Museum of Science, Technology and Transport)
NCO – Non-comissioned Officer
Oblt. – Oberleutnant
Oef – Oeffag
Offstv. – Offizierstellvertreter
Rb – Reihenbild
serg. – sergente
Stfw. – Stabsfeldwebel
ten. – tenente
U/C – unconfirmed
U/I - unidentified
WIA – wounded in action
Zgsf. – Zugsführer

Right: Aviatik D.I 338.53 in a hangar at Aspern. Several planes of this series served at Flik 74/J in the last months of the war but very little is known about their operational use. (Martin O'Connor via Aaron Weaver)